D1087826

# Air Quality in Cities

Springer

*Berlin*
*Heidelberg*
*New York*
*Hong Kong*
*London*
*Milan*
*Paris*
*Tokyo*

Nicolas Moussiopoulos (Ed.)

# Air Quality in Cities

With 127 Figures

Withdrawn
University of Waterloo

 Springer

Editor:

Dr.-Ing. Nicolas Moussiopoulos
Professor and Laboratory Director
Laboratory of Heat Transfer and Environmental Engineering
Box 483
Aristotle University
54124 Thessaloniki
Greece
E-mail: *moussio@eng.auth.gr*

ISBN 3-540-00842-x   Springer-Verlag Berlin Heidelberg New York

Library of Congress Cataloging-in-Publication Data Applied For

A catalog record for this book is available from the Library of Congress.
Bibliographic information published by Die Deutsche Bibliothek
Die Deutsche Bibliothek lists this publication in die Deutsche Nationalbibliographie; detailed
bibliographic data is available in the Internet at <http://dnb.ddb.de>.

This work is subject to copyright. All rights are reserved, whether the whole or part of the material is
concerned, specifically the rights of translation, reprinting, reuse of illustrations, recitations,
broadcasting, reproduction on microfilm or in any other way, and storage in data banks. Duplication of
this publication or parts thereof is permitted only under the provisions of the German Copyright Law
of September 9, 1965, in its current version, and permission for use must always be obtained from
Springer-Verlag. Violations are liable for prosecution under the German Copyright Law.

Springer-Verlag Berlin Heidelberg New York
a member of BertelsmannSpringer Science+Business Media GmbH

http://www.springer.de

© Springer-Verlag Berlin Heidelberg 2003
Printed in Germany

The use of general descriptive names, registered names, trademarks, etc. in this publication does not
imply, even in the absence of a specific statement, that such names are exempt from the relevant protective
laws and regulations and therefore free for general use.

Cover Design: oh! Holleschek Grafik Design, Garmisch Partenkirchen  and
Struve & Partner, Heidelberg
Typesetting: Camera-ready by Sofia Eleftheriadou

Printed on acid free paper    30/3140  – 5 4 3 2 1 0

# Preface

Urban areas are major sources of air pollution. Pollutant emissions affecting air quality in cities are considered to have adverse consequences for human health. Public and government concern about environmental issues arising from urban air pollution has increased over the last decades. The urban air pollution problem is widespread throughout the world and it is important to find ways of eliminating or at least reducing the risks for human health.

The fundamentals of the physical and chemical processes occurring during air pollutant transport in the atmosphere are nowadays understood to a large extent. In particular, modelling of such processes has experienced a remarkable growth in the last decades. Monitoring capabilities have also improved markedly in the most urban areas around the world. However, neither modelling nor monitoring can solve urban air pollution problems, as they are only a first step in improving useful information for future regulations. The defining of efficient control strategies cannot be achieved without a clear knowledge of the complete pollution process, i.e. emission, atmospheric transport and transformation, and deposition at the receptor.

Improving our ability to establish valid urban scale source-receptor relationships has been the objective of SATURN, one of the 14 subprojects of EUROTRAC-2. Similar to the other subprojects of this co-ordinated environmental project within the EUREKA initiative, SATURN brought together international groups of scientists to work on problems directly related to atmospheric chemistry and physics. The present volume summarises the scientific results of SATURN. The various chapters in the book reveal that research conducted in SATURN led to a significant state-of-the-art improvement in air pollution research.

A scientific synthesis such as the one attempted in the present volume has numerous contributors. As the co-ordinator of SATURN and the editor of its final report I would like to express my thanks: to the SATURN principal investigators who have carried out high-quality scientific work; to the chapter authors who all did an excellent job in presenting the major findings in a concise manner; to all other contributors to this final report who substantially supported the chapter authors; to the participating governments in EUROTRAC-2, all other national or regional funding agencies and the European Commission for funding our research (cf. overleaf); to Drs Pauline Migdley and Markus Reuther at the EUROTRAC-2 International Scientific Secretariat for their continuous support; to Dr. Petroula Louka, the SATURN Scientific Secretary, for her enthusiastic support in the co-ordination of SATURN and her decisive role in the preparation of the final report; and, finally, to Mr. Christian Witschell and his colleagues at Springer for providing the opportunity to present the results of SATURN in a way which will bring them to the notice of the large communities of atmospheric scientists and environmental managers.

On behalf of all authors and contributors to this report, but also in the name of the whole SATURN community I acknowledge the financial support of:

*The agencies:* Academy of Finland; ADEME, France; AgipPetroli, Italy; British Council; CESI, Italy; Hungarian Meteorological Service; City Authorities of Watford and Westminster, UK; Committee for Nature Use, Environmental Protection and Ecological Safety of St. Petersburg City Administration; Czech Ministry of Education, Youth and Sport; Danish Ministry of Environment; Danish National Research Council; Eastern Electricity plc, UK; Economics and Social Research Council, Go-Ahead Group plc, ETI Group, and Rupprecht and Patashnick, UK; Engineering and Physical Sciences Research Council, EPSRC, UK; Environment and Health Protection Administration of Stockholm; FCT, Portugal; Finnish Ministry of the Environment; Finnish Ministry of Trade and Communications; Finnish Technology Development Centre; French Ministry of Environment; German Ministry of Science and Education; German Federal Environmental Agency; Hellenic General Secretariat of Research and Technology; ICCTI, Portugal; INDRA S.A., Spain; Italian Environmental Agency, ANPA; Italian University and Research Ministry; Mexican Petroleum Institute; Ministry of Environment & Planning, Portugal; Ministry of Science & Technology, Portugal; Ministry of Science and Technology, Spain; National Road and Traffic Administration of Stockholm; Natural Environment Research Council, UK; PAS, France; Province of Brescia, Italy; Regione Lombardia, Italy; Russian Foundation for Basic Research; Swedish Environmental Protection Agency; Swedish Agency for Innovation Systems; Thai Government; University of Aveiro, Portugal; UK Engineering and Physical Science Research Council; UK Government Department of Environment, Food and Rural Affairs, DEFRA.

*The European Commission for the projects:* AIR-EIA (INFO 2000 Programme); APNEE and APNEE-TU (Information Society Technologies); IMMPACTE, CICYT REN2000-1754-C02-01/CLI and REN2000-1754-C02-01/CLI; IRENIE (Telematics for the Environment Programme); TRAPOS (European Training and Mobility Project), SUTRA-EVK4-CT-1999-00013.

*Other national funds:* Atmospheric Research 2000 program; Austrian Science Fund grants P 12168-TEC, P12170-TEC and P-14075 TEC; BWPLUS (Sustainability Research Programme of Baden-Wuerttemberg); CNRS programmes PATOM and PNCA; German Tropospheric Research programme; Finnish MOBILE and SIHTI projects; National Air Quality Monitoring Programme and Urban particles studies, Denmark; Czech project OE32 (E! 1489); Hellenic programmes of International cooperation with the UK, EUREKA (E! 1489), and YPER '97; Strategic Environmental Research Programme, Environment and transport (TRIP); French PRIMEQUAL-PREDIT programme.

SATURN's Framework Project was financially supported by the Shell Sustainable Energy Programme

Nicolas Moussiopoulos
Thessaloniki, January 2003

# Table of Contents

# List of Authors and Contributors

## Authors

Raimund Almbauer    Graz University of Technology, Austria
e-mail: almbauer@vkmb.tu-graz.ac.at

Carlos Borrego    University of Aveiro, Portugal
e-mail: borrego@ua.pt

László Bozó    Meteorological Service, Budapest, Hungary
e-mail: bozo@met.hu

Rex Britter    University of Cambridge, UK
e-mail: rb11@eng.cam.ac.uk

Ian Colbeck    University of Essex, Colchester, UK
e-mail: colbi@essex.ac.uk

Giovanna Finzi    University of Brescia, Italy
e-mail: finzi@bsing.ing.unibs.it

Stefano Galmarini    Joint Research Centre/European Commision, Italy
e-mail: stefano.galmarini@jrc.it

Jaakko Kukkonen    Finnish Meteorological Institute, Helsinki, Finland
e-mail: Jaakko.Kukkonen@fmi.fi

Steinar Larssen    NILU, Kjeller, Norway
e-mail: steinar.larssen@nilu.no

Petroula Louka    Aristotle University Thessaloniki, Greece
e-mail: petroula@aix.meng.auth.gr

Patrice Mestayer    Ecole Centrale de Nantes, France
e-mail: Patrice.Mestayer@ec-nantes.fr

Nicolas Moussiopoulos    Aristotle University Thessaloniki, Greece
e-mail: moussio@eng.auth.gr

Finn Palmgren    NERI, Roskilde, Denmark
e-mail: fpj@dmu.dk

Michael Schatzmann    University of Hamburg, Germany
e-mail: schatzmann@dkrz.de

Heinke Schlünzen    University of Hamburg, Germany
e-mail: schluenzen@dkrz.de

Ranjeet Sokhi    University of Herfordshire, Hatfield, UK
e-mail: r.s.sokhi@herts.ac.uk

Peter Sturm    Graz University of Technology, Austria
e-mail: sturm@vkmb.tu–graz.ac.at

Oxana Tchepel    University of Aveiro, Aveiro, Portugal
e-mail: oxana@dao.ua.pt

Dick Van den Hout    TNO/MEP, Apeldoorn, The Netherlands
e-mail: hout@mep.tno.nl

Marialuisa Volta    University of Brescia, Italy
e-mail: lvolta@bsing.ing.unibs.it

# Contributors

J.Amorim, University of Aveiro, Portugal
Y.Andersson-Skoeld, Melica Environmental Consultants, Goeteborg, Sweden
H.ApSimon, Imperial College London, UK
J.M.Baldasano, Univ. Politecnica de Catalunya, Barcelona, Spain
B.Calpini, Ecole Polytechnique Fédérale de Lausanne, Switzerland
D.Carruthers, Cambridge Environmental Research Consultants, UK
A.C.Carvalho, University of Aveiro, Portugal
A.Clappier, Ecole Polytechnique Federale de Lausanne, Switzerland
R.Colvile, Imperial College London, UK
A.Coppalle, Universite INSA de Rouen, France
A.M.Costa, University of Aveiro, Portugal
S.Despiau, LEPI, Universite Toulon-Var, La Garde cedex, France
S.Finardi, ARIANET s.r.l, Monza, Italy
E.Genikhovich, Main Geophysical Observatory, St. Petersburg, Russia
G.Graziani, Joint Research Centre/European Commission, Italy
O.Herbarth, Center for Environmental Research, Leipzig, Germany
O.Hertel, NERI, Roskilde, Denmark
C.Hirsch, Vrije Universiteit, Brussels, Belgium
M.Jicha, Brno University of Technology, Czech Republic
C.Johansson, Stockholm University, Sweden
K.Karatzas, Aristotle University Thessaloniki, Greece
B.Leitl, University of Hamburg, Germany
J.Levitin, Israel Meteorological Service, Bet Dagan, Israel
A. Lohmeyer, Ingenieurbüro Lohmeyer, Karlsruhe, Germany
M.Lopes, University of Aveiro, Portugal
C.Mensink, VITO, Mol, Belgium
A. I.Miranda, University of Aveiro, Portugal
A.Monteiro, University of Aveiro, Portugal
T.Pakkanen, Finnish Meteorological Institute, Helsinki, Finland
J.-P.Peneau, CNRS Physique et Images de la Ville, Nantes, France
S.Perego, Ecole Polytechnique Fédérale de Lausanne, Switzerland
P.Sahm, Aristotle University Thessaloniki, Greece
R.San José, Technical University of Madrid, Spain
E.Savory, University of Surrey, UK
T.Schoenemeyer, IFU, Garmisch-Partenkirchen, Germany
G.Smiatek, IFU, Garmisch-Partenkirchen, Germany
M.Sosonkin, Ukrainian National Academy of Science, Kiev, Ukraine
W.Winiwarter, Austrian Research Centre, Seibersdorf, Austria.

# Executive Summary

N. Moussiopoulos

Aristotle University Thessaloniki, Greece

## 1 Prologue

The main aim of SATURN has been to reach a better understanding of urban air pollution as a prerequisite for finding effective solutions to air quality problems and for sustainable development in the urban environment. To this end, the main scientific objective of SATURN has been to improve substantially our ability to establish source-receptor relationships on the urban scale. Ensuring the validity of such relationships may also facilitate the assessment of the impact of urban areas on regional and global scale problems of the atmospheric environment.

The following methodology was adopted to meet the above objective:

- Development of an appropriate model hierarchy, covering also the local scale (down to street canyon geometries) to the extent necessary to establish source-receptor relationships.
- Evaluation of individual models with suitable procedures.
- In support of such procedures, the creation of appropriate validation data sets from observations and experimental results originating from laboratory studies and field campaigns.

The structure of SATURN was based on three different clusters. The *Local Cluster* dealt with microscale and local scale phenomena investigated in the field, but also with wind tunnel experiments and numerical simulations. The *Urban Cluster* addressed urban-to-regional scale phenomena (without resolving individual obstacles) with field experimental campaigns and numerical models. Modelling work in both clusters included the development and validation of models with emphasis on the formulation of multi-scale concepts. The *Integration Cluster* concentrated on the integration of models in Air Quality Management Systems to be used for various applications including exposure estimates, and the testing/validation of such systems with collected data.

One of the major aims of SATURN was to contribute to the formulation of improved tools for urban air quality assessments. For instance, SATURN developed and improved methods to allocate the contributions of various sources to increased air pollution levels in conurbations. Such methods are very valuable for urban authorities wishing to have insight into the causes of air pollution and in the possibilities to reduce air pollution levels or to control anticipated increases of such levels. Also, methods to simulate the effect of long-term emission changes were improved as a basis for formulating and evaluating reasonable air pollution abatement strategies. On the same basis, novel systems for the prediction of air pollution episodes were among SATURN's deliverables. The modelling systems developed were validated with appropriate quality assurance techniques and together with the available technology, i.e. satellite information and telematics can be used for public information on air quality.

## 2 Characteristics of air pollution in European cities

A vast majority of the urban and suburban population in Europe is exposed to conditions that exceed air quality standards set by the World Health Organization (WHO). For exposure estimates it is necessary to investigate the spatial and temporal features of air pollution, especially under conditions that lead to the most severe air quality problems in different European regions.

Urban air pollution phenomena encompass a wide range of spatial and temporal scales: from a few metres (street canyon pollution) to hundreds of kilometers (secondary pollutant formation in city plumes). Within SATURN, special interest has been focused on stable, low wind conditions which lead to high pollutant concentrations. The effects of these meteorological conditions have been investigated in a local/street-canyon scale as well as in the larger mesoscale and regional scales. The microscale effects of low wind conditions have been investigated to determine the behaviour of pollutant dispersion in the vicinity of large ground-level sources. This is an important issue as an increasing number of roads in urban areas are constructed as cut and cover tunnels. Ground-level sources can be large parking areas, garages, small industries, etc. Other microscale phenomena that need special attention are those related to the heat fluxes in the vicinity of the buildings. Mesoscale wind circulations, e.g. mountain-valley wind systems, greatly affect the pollution dispersion in several industrialised and densely populated areas located in basins (e.g., in the area of the Alps). In addition, atmospheric circulations created by the city itself, notably the so-called urban heat island, directly influence the dispersion of pollutants. On the other hand, the local concentration levels are in many cases also influenced by regional scale processes such as the atmospheric transport of pollutants emitted in surrounding cities and industrialised areas as well as from neighbouring countries as in the case of ozone and particles.

The combined microscale, mesoscale and regional scale effects on urban air quality can be investigated with adequate multi-scale model cascades. The formulation of such cascades was among the main scientific objectives of SATURN. Several methods for coupling individual models were proposed in SATURN. This coupling allows consideration of the interactions between the various scales and assessment of the effect of such interactions on urban air quality.

An experimental campaign that helped to provide insight on the parameters affecting the airflow in a street canyon under low wind conditions was performed at the centre of Nantes, France, in June 1999. In particular, the effect of the thermal convection induced by the differential heating of the street surfaces due to variable solar heating on the airflow within the street and consequently on the pollution dispersion was investigated. The analysis of the air and wall temperatures showed the presence of a steep temperature gradient within the first few centimetres of the wall receiving direct solar radiation. This led to a strong convection which directly affected the transport of pollutants from the canyon to the layer aloft. The examination of the flow field indicated a main re-circulation during the whole day for wind perpendicular to the street. The thermal effects on the flow within a canopy were also investigated with wind tunnel experiments.

Investigation of dispersion under low wind conditions has been performed using a combination of mesoscale and Lagrangian particle models. In particular these models were applied for the Graz region in the lee-side of the Alps and used in conjunction with sonic anemometer observations. The results of the analyses made in a suburban area south of the city of Graz reveal peculiarities of the turbulent transport under low wind conditions influenced by the mixed land use and the low building density.

New modules of the French model SUBMESO were developed for simulating the mean flow and turbulence, the physics and microphysics, and the transport-diffusion-transformations of reactive pollutants within an urban area. The models include modules for dynamics, microphysics, terrain and soils, chemistry (transport-diffusion-transformation) fed by the tropospheric chemistry model MOCA, and a grid generator designed for complex terrain. The stable atmospheric boundary layer as well as urban canopy processes (radiative trapping, heat storage by the walls, mutual shadowing and vegetation over artificial surfaces) were investigated. Development and refinement of existing modelling tools were performed for the triangle Antwerp-Brussels-Gent, a highly urbanised area in northern Belgium, and integrated in the model system AURORA. The sensitivity to a set of variable input parameters (temperature, solar angle, etc.) was found to be high.

# 3 Methodology for specifying urban emissions

Emission inventories are important tools to describe the emission situation and eventually to manage air quality. They provide comprehensive information on emission sources and emission fluxes in the area under consideration. When looking at pollutant dispersion and air quality modelling, it is necessary to have an accurate and high quality description of the emission sources in terms of quantity and dynamic behaviour. This requires an emission inventory with a high resolution in space and time.

Consistency among urban emission inventories is a prerequisite for the sensible comparison of the air quality situation in various European cities. Within SATURN, guidelines were developed for setting up urban emission inventories to minimum requirements. The aim of these guidelines was to:

- present a harmonised method for the compilation of urban emission data,
- provide sources for suitable emission factors,
- provide tools for producing emission inventories.

For the generation of emission data, guidelines were designed for methodologies, models, pollutants, resolution in time and space, emission factors and quality assurance. In addition, a protocol for the comparison of urban emission inventories on qualitative and quantitative bases was established, and comparisons concerning pollutants treated and methodology used were performed.

The comparative analysis of several urban emission inventories suggests that it is not easy to find a common basis for intercomparison purposes. A first comparison was based on total emissions. The local emission situation differs strongly and is a function of the heating characteristics and the industrial activities in the domains under consideration. Special emphasis was put on the emissions from the industrial sector. In this sector the emission inventory is especially complicated as it is sometimes difficult to obtain information about the specific process activities in each industrial branch. Moreover, this kind of data accepts little aggregation.

A much higher potential for successful comparisons among various urban emission inventories was found for the road traffic emissions. The basis for the comparison is now the vehicle kilometres travelled, so that the result is a fleet averaged emission factor in grams per kilometre.

## 4 Contribution of urban emissions to the ozone budget

Photochemical pollution refers to the complex formation of ozone and other photochemical oxidants resulting from the interaction of high solar irradiation levels with the precursor substances of nitrogen oxides ($NO_x$) and hydrocarbons (VOCs). $NO_x$ is produced by combustion processes and is emitted by vehicles and industries mostly as NO with the balance being $NO_2$. VOCs originate from both anthropogenic and biogenic sources. Ozone is a major environmental concern because of its adverse impacts on human health. Since the 1970s health standards for ozone have been often violated in major metropolitan areas in Europe, especially in Southern Europe (e.g., Athens). The relative importance of ozone and its precursors transported into an area distinguishes the characteristics of an episode. Therefore, the needs to characterise the ozone episodes accurately and to identify the potential role of transport are closely related. As ozone is highly associated with the presence of $NO_x$ and VOCs, it is necessary to understand how ozone depends on these substances in order to develop an effective policy response.

Within SATURN, the OFIS (Ozone Fine Structure) model was applied to the investigation of the characteristics of photochemical pollution in an urban environment over periods of several months. The OFIS model belongs to the European Zooming Model system, a comprehensive model system for simulations of wind flow and pollutant transport and transformation. OFIS can provide a realistic statistical estimation of urban scale ozone levels and is therefore suitable for assessing ozone exposure at the urban scale for the optimisation of control strategies. The model was used for studying ozone formation in about 60 European cities in geographically different regions. The analysis of combinations of additional $NO_x$ vs. VOC emission reductions at the local scale has provided information on how to achieve maximum reductions of ozone levels in the conglomeration around each city or cluster of cities. In addition, an analysis was made as to how exposure to ozone is affected by regional and local measures associated with $NO_x$ and VOC emission reductions. One of the major results was that regional emission controls may have a significant influence on urban ozone concentration. Ozone exceedance statistics proved that the highest ozone values are present mainly in urban Mediterranean areas, while exceedances of the WHO guideline value for the protection of human health are also found in Central and Northern Europe, but only for shorter periods.

Two integrated modelling systems for photochemical pollution control were designed and applied to the Lombardia area in Italy. The modelling system STEM-FCM (Flexible Chemical Mechanism) was upgraded to describe the influence of different chemical mechanisms for ozone predictions and emission control strategies. The simulations made with the GAMES modelling system pointed out that, because of the main influence of the regional ozone production dynamics, the road network strategy has a modest influence on the concentration field. As for recent EU directives concerning vehicle technology, simulations show how pollution abatement measures addressing road traffic alone may prove inadequate. It must be stressed that the area studied is characterised by a VOC-limited atmospheric chemistry regime.

In addition, the experimental investigation of the chemical composition and aerodynamic distribution of the urban aerosol, performed over the Milan urban area in July 2000 during a typical summer high pollution episode, helped to understand which chemical multi-phase modelling schemes have to be implemented in the mesoscale models. Results mainly concerned the urban aerosol mass closure determination and the relation between secondary aerosol formation and photochemical properties of the atmosphere. Aerosol analysis was combined with a complete characterisation of the primary and photochemical pollutants, both at ground level and vertical structure.

Numerical modelling work concerning ozone formation dependent on $NO_x$-$O_3$ chemistry was also performed at the street-canyon scale with microscale modelling tools complemented by a module describing fast chemical reactions.

# 5 Particulate matter in European cities

Particulate matter (PM) is a suspension of solid and liquid particles in the air. PM is associated with adverse effects on health, climate and ecosystems. These effects are highly dependent on the size distribution and chemical composition of PM. Particulate matter may be either of primary or secondary origin. Those of primary origin are directly emitted by anthropogenic or biogenic sources, while those of secondary origin (the so-called secondary aerosols) are associated with long-range transboundary transport and are formed by atmospheric reactions arising from, e.g., emissions of $SO_2$, $NO_x$, $NH_3$ and organic compounds. SATURN aimed to investigate the chemical composition of size-fractionated aerosol in order to draw some general inferences about the sources and processes generating this class of pollutants in different European areas.

Several field experiments were conducted within SATURN in order to develop sufficiently detailed databases to enable source apportionment studies and characterisation of particles, including fine as well as ultra-fine particles. These experiments were performed in Stockholm, Toulon, Marseilles, Nantes, Helsinki, Copenhagen, Budapest and in a semi-rural site in the UK at Hatfield.

Traffic is considered to be one of the most important sources of fine particles. Determination of emission factors of fine particles from the actual car fleet for different types of vehicles is essential for reliable model calculations of the directly emitted particles from the traffic. Measurements of fine and ultra-fine particles were carried out during winter/spring 1999-2000 in Denmark, at street and roof levels in central Copenhagen, and at street level in the city of Odense. The major goal of the study was to provide tools for determination of traffic generated air pollution, and the collection of data for the development and validation of urban and local scale air quality models. Significant correlation at street level was observed between the CO, $NO_x$, and ultra-fine particles, indicating that the traffic is the major source of ultra-fine particles in the air. Time series of the number and mass size distributions were analysed over the period of several months using statistical methods. Factor analysis was used for identifying important sources, and a constrained physical receptor model was used for source apportionment and for determination of source size distributions of ultra-fine particles from diesel and petrol fuelled vehicles. Other pollutants related to traffic (e.g. benzene and toluene) were also monitored.

The physical characterisation of aerosol particles at ground and roof levels in the urban zone and their dynamic behaviour within the street were investigated with different experiments conducted in Toulon, Marseilles and Nantes. The measurements showed that the mass size distribution of fine particles corresponded closely to the distribution of traffic emissions at the field sites. Measurements of the coarse particle size distribution at several heights within the streets led to similar distributions characterised by a mode of approximately 2 μm. The fact that the concentrations were, on average, higher at the highest measuring level within the street than at street level suggests that the phenomenon of suspension of PM from the road dust deposit, which is believed to be the main process of coarse particle generation at street level, is not really dominant. On the other hand, the fact that the concentrations during rainy days were higher suggests that coarse particles are more commonly transported over the urban canopy and introduced into the road at roof level rather than generated at street level.

Gaseous pollutants and several quantities characterising PM were measured in an urban area and in a road tunnel in Stockholm. In detail, the measurements included aerosol size distribution, $PM_{2.5}$ and $PM_{10}$ mass (i.e. concentration of particles with size up to 2.5μm and 10μm, respectively), coarse and fine fraction elemental composition, organic and elemental carbon, CO, NO, $NO_2$, light hydrocarbons, semi-volatile hydrocarbons, polycyclic aromatic hydrocarbons, benzene, toluene, xylene and ethyl-benzene as well as filter samples. The hourly averaged total particle concentration reached its maximum in the road tunnel during morning rush hours. Most of the particles had aerodynamic diameters in the range 10 to 60 nm. The number of particles was lower in the street and the size distribution was shifted towards smaller particles. At rooftops the number of particles was found to be significantly lower. Some indication of formation of secondary particles was observed.

Between April 1996 and June 1997 concentrations of atmospheric particles and gases were monitored and size-segregated aerosol sampling and detailed chemical analysis combined with measurements of local weather conditions were performed at an urban and at a rural site in the Helsinki area. One of the aims was to estimate gas-particle interactions and particle deposition. More than 60 chemical components were determined and correlations between them were calculated. In the fine particle range four groups of correlating components were observed at both measuring sites. Average chemical composition of fine particles was fairly similar at the urban and rural sites.

The particle concentration was also investigated with field experiments in the UK. The results suggested that particle mass resides predominantly in the fine fraction ranges ($PM_1$) indicating their anthropogenic association. The results also indicate a low concentration of lead that might have resulted from the complete phasing out of leaded petrol in the UK. Generally the trend in size distribution exhibited by the analysed species is explainable in terms of their possible source origins. The size-distributed concentration results also suggest that most of the species have very high occurrence in the fraction of aerosols with diameter less than 1μm. These may have serious implications for human health and also indicate the extent of the influence of anthropogenic activities on the ambient aerosol.

Particle size distribution was also measured in the eastern Mediterranean area (in Crete and on a scientific boat located between the Greek mainland and the island of Crete). The first data analysis from this intensive measurement campaign indicates periods with elevated levels of number concentration and enhanced aerosol scattering.

# 6 Source apportionment and other modelling techniques

The chemical composition of atmospheric pollutants is influenced significantly by the nature of human activities leading to their emission. Therefore, the study of the source origin of the different elements is of crucial importance for environmental management on urban scale. Various studies within SATURN aimed at evaluating the characteristics of several trace elements of natural and anthropogenic origin under different conditions.

During a three year project (1998–2000) data were obtained in order to establish source-receptor relationships for hydrocarbons and particulate matter in the urban area of Stockholm, to evaluate the relative contribution of individual sources for the distribution of these compounds, and finally to evaluate emission inventories using a combination of measurements, source-receptor models and dispersion models. The project included both emission and air quality measurement campaigns and provided chemical and physical characterisation of the aerosol. The experimental results show that the particle number size distribution varies considerably with distance from traffic sources going from a road tunnel via streets with high traffic urban roof-top locations to a background site far from the city.

Particle interactions with local emissions and background air pollution may be characterised with suitable numerical methods. The assessment of existing and new methods of this kind proved this characterisation to be computationally nontrivial.

Receptor modelling was applied to provide quantitative estimates of the impacts of various sources on ambient air quality. Source profiles for waste incineration, traffic, oil and coal burning were established and used in model calculations. Aerosol sampling for fine size range aerosol particles was also carried out at the sources and at two receptor points, in the downtown area of Budapest during November-December 1999. Source signatures for coal and oil burning were adopted from the fine size range aerosol measurements carried out in the plume of power plants operating in the Czech Republic. It was found that high amount of zinc, lead and copper are emitted from a waste incinerator. Regarding the traffic profile, the most important element is still lead; however its relative contribution decreased rapidly during the past five years.

Mathematical models have also been developed, evaluated and applied in order to predict aerosol processes and the number and mass concentrations, such as $PM_{2.5}$ and $PM_{10}$, in urban areas. An aerosol dynamical model MONO32 has been applied both in Stockholm and Helsinki. The model takes into account gas-phase chemistry and aerosol dynamics (nucleation, coagulation, condensation/ evaporation and deposition). In Stockholm, the model was tested by comparison with measurements in a road traffic tunnel;  in Helsinki, the influence on aerosol evolution of various chemistry and aerosol processes was quantitatively evaluated.

Roadside emission and dispersion models for PM were developed and evaluated in the United Kingdom and Finland. The predictions of the CALINE4 and the CAR-FMI models were compared with field data. A simple model was developed for predicting the concentrations of $PM_{2.5}$ in urban areas, including the influence of both primary and non-exhaust vehicular emissions, and the regionally and long-range transported contributions.

Based on the observations that local vehicular traffic is responsible for a large fraction of the street level concentrations of both $PM_{10}$ and $NO_x$, either due to primary or secondary emissions (suspension from street surfaces), a semi-empirical modelling system for $PM_{10}$ was developed and evaluated against the data collected from the air quality monitoring network in the Helsinki Metropolitan Area.

## 7 Air pollution exposure – emphasis on particulates

It is unquestionable that air pollution has severe impacts on human health. In recent years increasing interest has been focused on aerosols, especially the particles of small diameter that are associated with epidemiological effects and respiratory diseases. In particular, a large number of health effects of traffic-generated air pollution are suspected. However, quantitative estimates that are required for risk assessment and management are extremely difficult due to the lack of data and/or difficulties with extrapolating data from other geographical areas and time periods with large differences in composition of air pollution. Moreover, very few and mainly small studies have so far attempted to assess individual exposures. The aspect of human exposure to ambient air pollution in urban areas has been extensively studied within SATURN. In particular, the relationships between particulate air pollution emissions and human/population exposure leading to potential effects on health have been investigated.

The complex concentration patterns in cities, particularly near hotspots such as roads, is a major reason why the relation between concentration and population exposure is far from trivial. Work was done on the assessment of road-user exposure to fine particulate air pollution in London using measurements and atmospheric dispersion modelling, specifically a comparison between the two. The road-user exposure with $PM_{2.5}$ was measured from the breathing zone for car, bus and bicycle users and compared with model results. Various human exposure studies were also carried out in Denmark and serve as the basis for model development and validation. Personal exposure to particles ($PM_{2.5}$) and nitrogen dioxide was monitored in Copenhagen, outdoors as well as indoors. Preliminary results indicate that a strong correlation exists between indoor and personal-monitored $PM_{2.5}$ concentrations indicating that indoor concentrations play a significant role for the personal exposure. The correlation between indoor and outdoor $PM_{2.5}$ concentrations seems less pronounced than expected, but more detailed analysis is required before conclusions may be drawn. The same study was aimed at extracting relationships between particle exposure and the response in biomarkers.

Mathematical transport-chemistry models are strong tools for the evaluation of emission reduction strategies, for providing information to the public, and as the central part of models for human exposure to air pollution. For instance, in Denmark, Finland, Norway and the United Kingdom, existing modelling systems have been extended to allow for the exposure of the population to air pollution. In Helsinki, the main objective was to evaluate the exposure of the population with a reasonable accuracy, instead of the personal exposures of specific individuals. Combination of the predicted concentrations, the location of the population and the time spent at home, at the workplace and at other places of activity was carried out. This modelling system has been designed to be utilised by municipal authorities in urban planning, e.g., for evaluating impacts of future traffic planning and land use scenarios.

Future scenarios of population exposure above the EU directive limits have also been investigated applying integrated software packages with operational functionalities suitable for use also by non-specialists. Several European cities have a $PM_{10}$ problem mainly due to the use of studded tyres on cars, and subsequent suspension of PM from the road dust deposit of worn away road surface. Applications included assessing various abatement strategies towards 2010 to reduce or eliminate the problem, as well as forecasting future $NO_2$ pollution levels.

## 8 Obstruction of local wind flow by buildings

Urban emissions occur mainly within or shortly above the urban canopy layer, i.e., within a zone where the atmospheric flow is largely disturbed by buildings and other obstacles. It is well known that, in comparison to unobstructed terrain, building effects can change local concentrations by more than an order of magnitude. In the street canyon scale, airflow and turbulent fields depend largely on the strongly non-linear interactions between street geometry and micrometeorology. As a consequence, the pollutant concentration fields within the streets and at roof level are poorly assessed. For this reason, the significance of sensors measuring proximity pollution levels within streets, e.g., for health impact assessment, is unknown, especially since vehicle pollutants transform rapidly during their dispersion within the streets themselves. In addition, the accurate representation of the airflow caused by the building obstruction is essential in order to adequately predict air quality in micro and local scales.

Wind tunnel and full-scale experiments have been conducted by SATURN groups in order to give a better insight into the airflow characteristics generated by obstacles, and to provide detailed databases for the evaluation of numerical model results. The effects of streamwise aspect ratio on the flow within a nominally two-dimensional street canyon whose axis is normal to the oncoming wind were studied in the wind tunnel of the University of Surrey. In particular, the width/depth ratios studied were 0.3, 0.5, 0.7, 1.0 and 2.0. The flow regimes have been studied experimentally using a range of instrumentation for wind velocity and turbulence measurements. The experiment showed that the flow regimes change from a single main re-circulation for aspect ratios 1.0 and 2.0, and a two-vortex flow for aspect ratios 0.5 and 0.7 to a very weak flow within the street canyon that is difficult to be measured by the instruments, indicating the importance of geometry on street ventilation.

The effects of building geometry on the flow become more complex when treating an urban quarter in three dimensions. The experiments conducted in the wind tunnel of the Hamburg University showed that even small changes in street canyon geometry can change the flow pattern within the canopy layer quite significantly. This was demonstrated in the example of the Goettinger Strasse in Hanover, Germany, for which field data was also available. Three different obstacle configurations were replicated in the wind tunnel. It was found that geometrical variations can fundamentally change the flow behaviour within street canyons. Adding geometrical detail or further buildings to the wind tunnel model modifies not only the flow inside the street canyon but also in the side streets. The flow in the streets perpendicular to the canyon can even change direction which results in a completely different ventilation pattern and thus large differences in local pollutant concentrations within the canyon. The airflow and pollution dispersion in this complex urban configuration have been simulated with the prognostic numerical model MITRAS. It was shown that the flow pattern is influenced by local obstacles or openings which should be considered when simulating pollutant transport at a local scale.

The collected data sets from all wind tunnel experiments were used to validate several computational fluid dynamics (CFD) codes, all applying the standard k-ε model. This exercise showed that the implementation of the wall function by the codes is mainly responsible for the differences between experiment and predictions that appear close to solid boundaries for the simple two-dimensional cavities. The results of a complex three-dimensional case showed discrepancies in distinct features of the flow patterns. On the other hand, comparison of model results with the field data illustrated that the agreement or disagreement found is not necessarily a quality indicator for the model when the measurements are taken at only one position within a street canyon close to building irregularities (as is usually the case). Due to the large variations in local concentrations, agreement and disagreement is likely to be random. Special attention should be paid to this result when decisions for the selection of positions for urban monitoring stations are taken. In the frame of SATURN it could be demonstrated that data collected under carefully controlled conditions in wind tunnels are useful for the proper validation of microscale numerical models developed for the simulation of dispersion processes within the urban canopy layer.

The Nantes '99 experimental campaign offered a detailed database concerning the turbulent characteristics of the flow arising from the presence of vehicles within the street. The analysis of the turbulence produced by the vehicle motion and its effect on pollution dispersion showed the influence of turbulent kinetic energy (TKE) produced by traffic. In low wind conditions, TKE increases up to a threshold value of vehicles density per lane, while above this number the vehicles act rather as a "moving wall", leading to a TKE reduction. The TKE was found to be correlated with the CO concentration within the canyon and, consequently, it was demonstrated that turbulence induced by traffic does influence pollution dispersion within the street. A method accounting for the influence of moving vehicles in specific urban situations, such as the area around road tunnel openings, street canyons and intersections has been established in CFD modelling, in order to predict the effects of traffic on the local scale pollutant dispersion.

The three-dimensional structure of the turbulent flow in the roughness sub-layer over an urban-type surface comprising regular arrays of roughness elements was also investigated within SATURN. Experimental study in the University of Surrey wind tunnel determined the growth of the depths of the inertial and roughness sub-layers over a flat surface covered with a regular array of 3D roughness elements. Spatial averaging was used to remove the variability of the flow due to the individual obstacles. This showed that the spatially averaged mean velocity in both sub-layers could be described by a standard logarithmic law, with mean zero-plane displacement, using the true roughness length and surface friction velocity.

## 9 Remote sensing and urban air quality management

In recent years remote sensing techniques have gained much importance in the context of air quality management. At the urban scale such techniques range from new monitoring approaches to the use of satellite information.

In the frame of SATURN, the Differential Optical Absorption Spectroscopy (DOAS) technique was applied for the analysis of air pollution in St. Petersburg. In addition, Fourier spectrometers were utilised to assess air pollution in Kiev. Furthermore, an interferometric method was applied for deriving roughness length information from satellite data. The knowledge of the appropriate roughness length is a prerequisite for the reliable description of transport velocity and deposition rates of atmospheric pollutants. This parameter is, therefore, required for environmental pollutant transport and deposition models. Until now roughness lengths were mapped on the basis of land use data from optical remote sensing systems. The disadvantage of this method is the uncertainty in the image classification and the high cost of data processing.

A mapping procedure for roughness length from synthetic ERS interferometry data has been developed within SATURN. Several tests with the real mode airborne Synthetic Aperture Radar (SAR) interferometry data were performed for an area near Bonn and the surroundings of Oberpfaffenhofen, Germany. The major aim of the tests, performed for different areas with different land use types, was to investigate if the typical scattering of the relative phase measurement found in the synthetic interferometry is also present in the real mode interferometry. The results show a high correlation between the standard deviation of the relative phase and the roughness for all major land use categories with the exception of waters.

# 10 Model Quality Assurance

In order for urban air quality assessments to be reliable, the models used should be quality assured. Model quality assurance (QA) procedures are designed to ensure that the appropriate methods and data are used, that errors in calculations and measurements are minimised and that documentation is adequate to meet the project objectives. As one of the subproject's main objectives, activities were performed within SATURN aiming at the evaluation of local and urban scale models with adequate data sets.

Urban air pollution model evaluation has to be based on an appropriate evaluation procedure and requires the availability of appropriate quality assured data from field studies and/or laboratory experiments. Detailed specifications are normally formulated for all data involved in evaluation procedures in order to avoid model evaluation attempts failing because of missing or inadequate data.

Within SATURN a new methodology was established for the validation of local scale models. This methodology is based on the utilisation of high quality data from wind tunnel measurements. A CFD model evaluation was performed through an intercomparison exercise leading to the conclusion that the implementation of the wall function plays an important role for the simulated flow in the vicinity of solid boundaries.

New knowledge regarding quality assurance of urban scale models resulted from intercomparison activities and sensitivity studies. Model intercomparison is very useful for identifying inconsistencies and for estimating model uncertainties. In particular, the MESOCOM exercise was conducted as a pilot study aiming to improve the transparency of urban scale models. The model results collected were analysed and published on the Internet. A major finding was that large differences exist among the various models, which were found to be primarily due to the surface layer parameterisation and the boundary condition treatment.

Six research groups participated in ESCOMPTE_INT. This model validation exercise aimed at assessing to what extent current models are capable of describing airflow and dispersion phenomena in the atmosphere above a coastal urban area. The exercise elucidated the strengths and weaknesses of the approaches adopted in the models and led to interesting conclusions regarding future research needs.

The use of experimental data in the model validation process makes the implementation of QA measures for data collection necessary, the principal objective being to quantify and to reduce the uncertainties of measurements. For this reason, Data Quality Indicators and Data Quality Objectives were used for the determination of the completeness and the range of acceptability of the data sets.

## 11 Urban Air Quality Management Systems

The European Air Quality Framework Directive and associated Daughter Directives require projections of air quality in urban areas for assessment against air quality objectives. In addition, air quality strategies necessitate the development of plans to reduce pollution in specific areas. The principal aim of the ongoing development is a practical management system, which can be used to undertake these tasks. This requires an advanced operational air quality model such that a range of air quality management scenarios may easily be assessed. These can include traffic management options such as low emission zones, clear zones or multiple occupancy vehicles etc. and other emission reduction options for other sources. Such analysis also requires that input parameters such as emission factors may easily be modified.

Products of policy relevance that resulted from SATURN work are related to the development and application of integrated systems for supporting urban air quality management as well as reliable tools for forecasting the frequency and severity of air quality limit value exceedances in cities. Therefore, the main common target of the SATURN groups working on such integrated systems was to extrapolate the present air quality into the future for various abatement scenarios. First attempts were performed towards linking Urban Air Quality Management Systems with the optimisation of abatement action packages (e.g., according to cost benefit calculations), using the AIRQUIS integrated air quality management system. The URBIS system was developed as a comprehensive tool for the assessment of air quality in conformity with the new EU air quality directives. The need for comprehensive assessment, both in space (full city) and in time (full year) in combination with high resolution in space (hotspot quantification) and time (hourly concentrations) renders it impossible to rely solely on full-scale 3D grid models. The ADMS-Urban system has undergone significant progress and testing with the development of the Atmospheric Emissions Toolkit, while the system was further used for many local authorities and several major applications regarding air quality review and assessment. The multi-scale model ZEUS was the basis for the development of OPUS-AIR, which is a policy-oriented system for the integrated assessment of technical and non-technical measures that are put forward in order to reduce urban air pollution levels. Several experiments have been performed with OPANA, a tool that may be used by city environmental offices in Europe to "validate" the air quality modelling tools they are using or are intending to use in the future.

Gradually, the tools developed within Urban Air Quality Management Systems for modelling and predicting air pollution are improving in quality and efficiency. At the same time more provisions have been built that link to elements that are at or even beyond the borderline of air pollution research: to the behaviour of the population and the related exposure and health risks; to the costs and the benefits of technical and non-technical measures; to other environmental topics that have strong relations with air pollution (noise, external safety).

## 12 Public information on air quality

One of the main objectives of SATURN was to investigate the key points in urban environment decision and policy making that interact with the environmental information produced and with relevant integrated environmental surveillance, information and management systems. In order to meet these aspects, activities focused on the requirements for the design and development of such systems, the development of air pollution forecasting tools and the scientific support of multimedia presentation and guidance related to environmental impact assessment related information.

Urban Air Quality Management and Information Systems are systems produced for sophisticated, integrated applications that are city-tailored. Therefore, there are no standards for their development and construction. The development process (life cycle) of such systems results mostly from experience. Also their content may vary significantly from one application to the other. Within SATURN, air quality forecasting tools were developed and tested against hourly mean values of air pollution data in different European urban areas. They were based on time-series models that may be applied for short-term predictions of urban air pollutants such as ozone and $SO_2$. Moreover, statistical techniques were used as a tool in the development and implementation of a warning system for personal exposure that is designed to support the air quality management of local authorities. A statistical tool based on the Classification and Regression Tree analysis was also used, in particular for ozone forecasting.

Another important application within SATURN was the use of multimedia for an effective interpretation of environmental information. A CD-ROM application was developed with the aim to instruct those wishing to learn more on methods and procedures regarding the compilation of an Environmental Impact Assessment study. The application also includes a general introduction on the basis of a carefully planned presentation scenario.

## 13 Epilogue

SATURN combined 40 individual contributions from 19 European countries and the European Commission. The total costs of the research conducted is of the order of € 50 million. The results of all contributions were integrated in order to be used for air quality management. The most important prerequisite for bridging the gap between environmental research and management are regular interactions between environmental scientists and decision makers. For this reason, SATURN developed a common structure for the scientific results (data, models, methods) providing an application-oriented framework to be used by the researchers as well as allowing for monitoring the performance of the contributions taken together and analysing the results in terms of applicability. As a result, novel modelling systems were developed comprising models and tools that are properly validated by specially developed techniques. These systems are already being used by local authorities for air quality management and population information.

The advances in the science of urban air pollution achieved within SATURN may be summarised as follows:

- Contributions of field studies to a better insight into the characteristics of polluted urban air
- Significant refinement of urban scale models and methods used therein; valuable new knowledge regarding quality assurance of these models resulting from intercomparison activities and sensitivity studies
- Substantial progress in the development and application of local scale models and establishment of a methodology for their validation
- Development of versatile Urban Air Quality Management Systems and their installation for use by city authorities; successful integration of urban air pollution research into the areas of information and communication technology

Despite the scientific progress achieved in SATURN, several issues related to urban air pollution have not yet been fully explored. More knowledge is needed on the pollutant emissions due to the real traffic behaviour and the sources of $PM_{10}$ and $PM_{2.5}$ affecting the urban atmosphere. Further research is also needed on the chemical and physical properties of particles with emphasis on the formation and regional transport of secondary aerosol. Novel multi-scale model systems based on nested domains with different models will be needed for assessing the impact of urban sources vs. long-range transport with regard to air quality. Also, more effort will be required for the accurate prediction of pollutant concentrations at hotspot locations. From today's perspective, this aim can only be accomplished with sophisticated local scale models. These, however, will have to be better adapted to regulatory needs. In this context, the combined and fully integrated use of measuring and modelling for air quality assessments is still to be promoted. Finally, the further development of standardised Quality Assurance procedures will result in an increased quantitative level of confidence of model predictions. More and higher quality data from field campaigns and laboratory experiments will be required to meet the needs of complete model evaluation procedures.

# Chapter 1: Introduction

N. Moussiopoulos

Aristotle University Thessaloniki, Greece

In the past few years a substantial body of scientific knowledge on air pollution has been built up, to a large extent in the framework of large international research activities. As a result, there are now many national emission databases and model systems that, at least partially, suit the environmental management needs of national and international authorities. The level of aggregation of these data and systems tends, however, to be too high for addressing urban air quality issues. At the same time, air quality in several European cities is far from acceptable and often above the limit values. Human activities lead to changes in the composition of the urban atmosphere that are considered to have serious effects on human health and ecosystems, while causing damage to materials and monuments. At present there is much public concern over high concentrations of photochemical pollutants and particulate matter in numerous conurbations all over Europe.

In the light of the above, SATURN was launched as the "urban" subproject of EUROTRAC-2, the Eureka project on the Transport and Chemical Transformation of Environmentally Relevant Trace Constituents in the Troposphere over Europe (second phase, 1996-2002). SATURN was accepted as a Eureka project in February 1996, and was concluded at the end of December 2002. SATURN fitted very well into the main goal of EUROTRAC, i.e. to improve the quantitative understanding of the factors determining the formation, transport, chemical transformation, deposition and impact of photo-oxidants, aerosols and other trace substances in the atmosphere, and in this way to contribute directly to the further development of strategies for reducing the anthropogenic contribution of these species.

The main objective of SATURN was to substantially improve our ability of establishing source-receptor relationships at the urban scale. Ensuring the validity of such relationships may also facilitate assessing the impact of urban areas on regional and global scale problems of the atmospheric environment. The following methodology was adopted for meeting the above objective:

• Development of an appropriate model hierarchy, covering also the sub-urban (local) scales to the extent necessary to establish source-receptor relationships.

- Evaluation of individual models with suitable procedures.

- In support of such procedures, the creation of proper validation datasets from observations and experimental results originating from laboratory studies and field campaigns.

SATURN's work programme was formulated with the intention of combining innovative research with substantial practical contributions towards more effective urban air quality management. Along these lines, key subproject products were to include novel systems for a comprehensive overview of source-receptor relationships both for peak and city background levels of air pollution. In spite of their small spatial extent (e.g., a few meters from traffic), peaks are important because they give the largest problems and also exceed limit values most frequently. City background levels, on the other hand, are associated with the exposure of the general population in the city. From the view of measuring concentrations, covering both peak and city background levels implies that measurements need to account for so-called 'hot spots', as well as to allow extrapolation to general concentration patterns in the city. Concerning the necessary modelling tools, it was clear from the outset that adequate coupling of urban and local scale models was required.

Within SATURN care was taken to improve our understanding of several scientific issues closely related to air pollution at local and urban scales. Such issues include:

- Urban emissions occur mainly within or shortly above the canopy layer, i.e., within a zone where the atmospheric flow is heavily disturbed by buildings and other obstructions. This may affect local concentrations by more than one order of magnitude compared to unobstructed terrain. Consequently, considering the buildings not individually but only within the roughness parameterisation may lead to significant errors in model simulations. SATURN aimed at the development of new types of high-resolution models and the creation of data sets suitable to evaluate these new models.

- The receptor points are close to the sources. Due to the combined effect of natural wind variability and obstacles, concentrations differ substantially over relatively short distances. The presence of large concentration gradients makes it difficult to find representative locations for air quality monitoring stations. Within SATURN methods should be developed for the identification of suitable measurement positions and for the spatial generalisation of point-wise measured data.

- Concentrations measured in urban environments are usually highly intermittent. Peak concentrations can be much higher than the corresponding mean values. One of SATURN's goals was to formulate specifications for measurement strategies and necessary equipment for operation under non-stationary conditions that are characteristic of the urban environment.

The major scientific challenge in SATURN has been to account properly for the ensemble of physical and chemical processes that affect urban air quality and that occur over a wide range of space and time scales (cf. Chapter 2). Towards this aim, in the early days of SATURN individual "main groups of activities" (experiments, model development, model evaluation) were distinguished in the subproject structure, while activities in each "main group" were organised in tasks according to their scale[1].

In the intermediate phase of the subproject and with the aim to maximize synergy among scientific groups active in the individual "main groups", activities in SATURN were re-grouped into the two newly introduced "clusters". The local cluster dealt with microscale and local-scale phenomena that are investigated using wind tunnel and field experiments as well as with numerical models. The urban cluster tackled urban-to-regional scale matters (without resolving individual obstacles) with field experimental campaigns and numerical models. Both clusters included the development and validation of models and novel modules.

In addition to the above, SATURN contained (initially in a "main group" and later in its own "cluster")a range of activities related to the integration of data, models and other tools required for effective urban air quality management.

The success of SATURN is closely linked to its inter-disciplinary character, in line with the overlay strategy of EUROTRAC-2: Prior to SATURN there was extensive research throughout Europe on meteorology, pollutant emissions, dispersion and air quality in the urban context. From the research co-ordination point of view, the main innovation in SATURN was to provide a strategic forum for the scientific and operational communities to collaborate, compare approaches and share developments. More than 40 scientific groups from 17 European countries and the European Commission participated in SATURN. As all EUROTRAC-2 subprojects, SATURN adopted the "bottom-up" philosophy, i.e. all contributions were funded by national bodies or by international organisations. The total research resources brought together by the participating countries and the European Commission were of the order of 50 million €.

SATURN has been a scientific platform covering all Europe and at the same time a training facility for young students. A European network of urban air pollution scientists has been formed, qualified to advise their own governments and local authorities on issues related to urban air quality. The numerous references in the various chapters of this final scientific report of the subproject reflect an extraordinary scientific output of the SATURN investigators over the last six years. There have been several hundreds of papers in the peer-reviewed scientific literature, as well as contributions to the Symposium of EUROTRAC-2. In addi-

---

[1] According to the terminology adopted in SATURN, in the *local scale* buildings and other obstructions in the urban canopy are being directly resolved, while in the *urban scale* they are described with suitable parameterisations, i.e. mathematical formulae linking characteristic properties of the urban canopy to other key model variables.

tion, the SATURN Annual Reports prepared yearly between 1998 and 2001[2] contain contributions submitted by the individual investigators.

Overall, the subproject contributed to a significant improvement of the state-of-the-art in the science of urban air pollution, which is summarised in the present final scientific report of the subproject. The new knowledge gained in SATURN provides the basis for effective strategies towards cleaner air in European cities.

The structure of this book follows the main intention to include all major scientific results of SATURN while maintaining a logical and coherent route. In all chapters the attempt is made to account for the interaction of all scales associated with urban air pollution.

The various phenomena occurring at the individual scales are discussed in Chapter 2 and aspects related to air pollutant emissions are outlined in Chapter 3. The subsequent Chapter 4 includes the results of the experimental campaigns performed within SATURN, both at the local and the urban scales, while Chapter 5 contains the significant new knowledge acquired with regard to particulate matter in urban air.

Important new developments concerning air pollution models, both at individual scales and in multiscale cascades, are the subject of Chapter 6. The state-of-the-art regarding Model Quality Assurance progressed singificantly within SATURN. Chapter 7 deals both with conceptual improvements and new approaches for the evaluation of urban and local scale models.

Following recommendations expressed on the occasion of the subproject's mid-term review, a major focus of SATURN has been to advance our knowledge on photochemical pollution in South European cities. Chapter 8 summarises the current understanding on urban photochemistry in South Europe as reflected in the scientific results of various SATURN groups and other European projects. Finally, Chapter 9 reveals that SATURN contributed substantially to the improvement of integrated urban air quality assessment and to the formulation of versatile tools for air quality management.

Hundreds of illustrations are included in the various chapters of this book, a large part of them presenting the results of complex mathematical model simulations. Coloured versions of the illustrations together with selected computer animations and additional information on SATURN can be found in the Internet (URL 1.1).

---

[2] References to the Annual Reports are not included in the bibliographic list given at the end of the report for all chapters; the insertion in the text corresponding to the contribution of Smith to the SATURN Annual Report 1999 reads (Smith ✑1999)

# Chapter 2: Urban Air Pollution Phenomenology

R. Britter

Department of Engineering, University of Cambridge,Cambridge,CB4 1HW,UK

## 2.1 Space and time scales in the urban context

Urban air pollution involves physical and chemical processes over a wide range of space and time scales. Cities have typical spatial scales up to 10 to 20 km, or possibly even larger when several adjacent urban areas coalesce to form a conurbation. Pollutant dispersion from near-ground level sources would be present through much of the atmospheric boundary layer over these distances. Processes at the city scale influence the larger regional scale up to 100 or 200 km by providing a momentum sink and a thermal and pollutant source. At the same time the regional, or larger, scale physical processes provide the background state for the city scale processes. Ozone episodes occur over the entire city or larger conurbation and, as they occur under particular meteorological conditions, may occur over a much larger region.

The city contains its own inhomogeneities, frequently with a city centre having larger buildings and an outlying area of lower industrial or residential areas as in Figure 2.1 of a section of Birmingham, UK. Some researchers find it useful to consider the city as consisting of many semi-homogeneous neighbourhoods of typical scale 1 to 2 km. Over these distances the dispersion processes may be influenced by the buildings themselves and the lower part of the atmospheric boundary layer, called the roughness sublayer that extends up to 2 or 3 times the average building height.

Peak pollutant concentrations will obviously occur when the pollutant source and receptor are nearby, for example when the receptor (an individual or a pollutant monitoring station) is at roadside and the pollutant source (vehicles, at high traffic density) is also within the street. Consequently processes at the street (or street canyon) scale, that may be up to 100 or 200m, are of particular interest. At this scale the geometric arrangement of the buildings, street canyons and intersections directly affect the dispersion processes.

**Fig. 2.1.** Part of the city of Birmingham, UK

There are also processes of interest at the "tailpipe" scale where the formation or growth of ultrafine (sometimes nano-) particles can be influenced by the details of the mixing process between the exhaust gas and the atmosphere.

For each spatial scale there is a corresponding time scale that is the spatial scale divided by a representative advective velocity, typically the wind speed. The flow and dispersion processes are physically much the same at all the scales although buoyancy or density driven flows will manifest themselves more easily if applied over large spatial scales e.g. the heat island over a city (Oke 1987).

The different spatial scales show themselves more clearly when one considers how to develop models for flow and dispersion. It is natural that the larger the spatial scale under study the less the importance and hence necessity for any detailed information. Alternatively the limitations on analytical or computational resources will not allow the detail to be studied over a large spatial scale. A further way of addressing this issue is to consider the concentration at a point receptor within a city. The concentration or concentration-time history will have a contribution from the background concentration arising on a regional scale and this will vary slowly and relatively smoothly with time. There will also be a contribution from the pollutant sources upwind of the receptor. The variations of the source positions and strengths will be smoothed out to a greater extent when well upwind and to a lesser extent for closer sources. (Moving) sources within the same street and near the receptor will produce rapidly varying and high magnitude concentrations. Thus the degree of detail required to address each scale reduces as the scale itself increases.

Very similar effects are seen when the concentration-time history at the point receptor is put through a running time-average of increasing time scale or volume averaging is used rather than a point receptor. In the latter case the variations with time will be reduced as the averaging volume is increased. Regulations are typically couched in terms of both temporal averages (15 minutes, 1 hour, 24 hour, annual) and, more recently, in terms of measuring volumes or spatial averages (an average over 200 sq. metres has been mentioned). These considerations are understood and are often included, sometimes implicitly, in urban pollution studies.

A somewhat more difficult problem is presented when chemical reactions are considered. The reactions will have their own chemical time scales either for individual reactions or for situations that involve very many chemical reactions each with their own time scale. The reaction relies on molecules being in proximity to other molecules and therefore on the mixing process between the pollutants and their environment. And the mixing process has its own time scale. The outcome of reactive flows will depend upon the relative time scales for the chemistry and for the mixing processes. Thus particular aspects of urban chemistry are more or less important dependent upon the spatial or temporal scale under study.

Urban air quality studies or predictions are directly dependent on the magnitude and location of the pollutant sources. Much effort is expended in developing emissions inventories for cities, preferably in a common format (McInnes 1996). These inventories will also be subject to selected temporal and spatial scales. The temporal scale is typically 1 hour. For regional problems a resolution of 1km by 1km would be appropriate while for the city scale a linear resolution of 250m by 250m would be satisfactory. For the street canyon scale the resolution is obviously based on the street itself. Emissions inventories are commonly based on generalised emission factors combined with information specific to a particular study.

The predictions of annual average $NO_2$ concentrations in London (Fig. 2.2) show a range of spatial scales. The main streets and intersections are clearly visible as a detailed filament-like array within the larger city-scale. To the west of London, Heathrow airport could be interpreted as a type of neighbourhood. Of course here we are seeing a pollutant concentration map rather than a change in the urban form. High concentrations are seen near the streets, reducing quite rapidly away from the streets. In central London the proximity of the streets to one another leads to the general background level over the whole area being high.

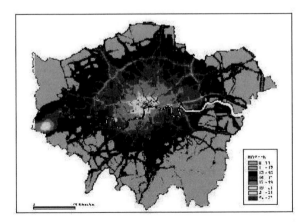

**Fig. 2.2.** Predictions of annual mean $NO_2$ concentrations for London, UK for 2005. With permission of CERC Ltd

## 2.2 Flow, dispersion and chemistry / street and receptor scales

The street scale is of particular interest for urban air quality as it is the smallest scale that encompasses one of the main pollutant sources, the vehicles and an important receptor, the public, and thus is likely to be the scale that determines the extreme values of concentration.

The flow on this scale is typically viewed as a recirculating eddy, as in Figure 2.3, within the street canyon driven by the wind flow at the top with a shear layer separating the above-canyon flow from that within (Hosker 1985). When the street canyon is relatively deep the primary vortex will not extend to the ground and there may be a weak contra-rotating vortex near the ground. The primary vortex may not extend completely across the street if the street canyon is relatively shallow. There are variations on this flow due to wind directions not normal to the street axis, the real rather than the simple idealised geometry of the street canyon and the mean flow and turbulence generated by vehicles within the street canyon (Pavageau et al. 2001).

Pollutant emissions from vehicles will dilute as they exit the vehicle exhaust initially through a jet-like mixing process, then by the turbulence in the vehicle wake, then by the turbulence in the street canyon as a whole. The recirculating eddy will advect the pollutants released near ground level towards the lee wall and away from the downwind wall. Thus higher concentrations are anticipated and observed on the lee wall. Maximum concentrations at street side receptors occur when the wind is either normal, or parallel, to the wind direction above the canyon. Whether the flow is best considered as an organised eddy structure with some

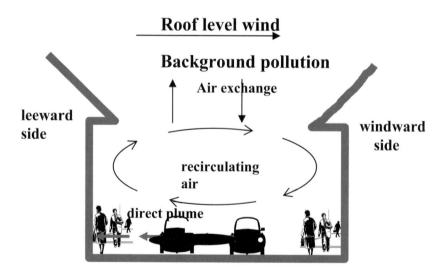

**Fig. 2.3.** A schematic of the flow in a street canyon

turbulence superimposed or as a strongly unsteady flow whose time-averaged flow pattern is a recirculating eddy is sometimes queried. Very idealised laboratory experiments can be distinctly different to real atmospheric flows that have large natural variability. There is also the possibility of thermally driven flows arising from solar insolation, building heat sources and from the vehicles themselves. These may disrupt the recirculating eddy structure.

The average level of the pollutant concentrations within the street canyon is a balance between the emissions within the street canyon, and the exchange of pollutants between the street canyon and the flow above, or possibly with other streets. The exchange is a two-way process with street-sourced material being removed and material from other sources upwind being brought into the street. The process can be parameterised with an exchange velocity. The magnitude of this exchange velocity has not been measured directly though it does appear in some operational street canyon models.

Little is known about the importance of any stabilising density stratification within the street canyon. Of course traffic flow rates at levels of concern for air quality should be more than adequate to remove any density stratification.

One particularly significant scenario is that of a highway tunnel exit. Here the pollutant source will be a large area source (e.g. the tunnel cross-section) with momentum and buoyancy that will interact with the ambient flow. The highway tunnel exit will aggregate sources that would otherwise be distributed along a section of road and place them in a very localised area.

Flow and dispersion at the street canyon scale are amenable to computational fluid dynamics though simpler models are more commonly used for operational purposes. Wind tunnel modelling provides well controlled experiments to determine the flow physics and dispersion while field studies ensure that the modelling is an adequate representation of the reality. There is uncertainty about the level of detail that must be modelled for both computational fluid dynamics and wind tunnel modelling.

There is also some uncertainty as to how best to "reference" the street-canyon problem. The reference often used is a roof height measurement and this is somewhat vague. Without a clear "reference" it can be difficult to generalise a "street-canyon" flow/dispersion from one site-specific study.

The prediction, on the street scale, of pollutant concentrations will be very sensitive to the assumed distribution of the pollutant source. The pollutant source can be temporally averaged over whatever is the relevant regulatory averaging time and this is typically an hour or longer. The spatial averaging is commonly to spread the source evenly along a length of the street. This may be a little too simple. Intersections, traffic lights etc. causes "stop-start" traffic patterns with the likelihood of large emissions locally due to both idling and acceleration.

At the street scale it is normally assumed that reactions that are fast compared with the time scale for ventilation of the street canyon occur while slower reactions are not considered. This leads to several quite simple chemistry mechanisms.

The principal interest is the conversion of NO from the vehicles to $NO_2$. Whatever the mechanism there is a need for some information regarding pollutant levels being introduced into the canyon to form a chemical background, $O_3$ in particular. The chemistry is generally taken to occur independent of the dynamics of the flow and the detailed mixing processes between the emissions and the airflow in the canyon. Simple operational models, incorporating empirical constants, may obscure any errors arising due to poor treatment of the mixing and chemistry. The chemical mix that leaves the street canyon will be the source term for the flow above and could warrant a more thorough treatment of the mixing and chemistry processes within the canyon. Shading within the street canyon reduces photolysis.

Particulate matter, $PM_{10}$, $PM_{2.5}$ and ultra-fine particles are all prevalent within the street canyon (Wåhlin et al. 2001). These may be directly emitted by the vehicles or other sources, be introduced into the street canyon with the ambient air from above, be deposited to and resuspended from the roadway, or grow as secondary parrticulates within the street canyon. Coarse particles and ultra-fine particles tend to remain close to their source due to gravitational deposition and diffusion to the surface respectively. Intermediate size particles remain suspended over greater distances.

It is not just the measured or predicted pollutant concentration field that is of interest. The exposure of a typical individual or the population at large to the pollutants is the real goal of urban air pollution studies (Adams et al. 2001). The link between the concentration field and exposure can be quantified through population density maps alone or by incorporating information on typical population daily movements. A related concern is the representativeness of regulatory monitoring sites (Scaperdas and Colville 1995).

Within SATURN field experiments were conducted for the identification of important pollution sources with emphasis on particulate matter. Wind tunnel experiments resulted in further insight into the complex flow and dispersion phenomena in street canyons and other characteristic local scale geometries. Computational methods were successfully applied for quantitative reproductions of local scale air flow phenomena.

## 2.3 Flow, dispersion and chemistry / neighbourhood and city scales

On the city scale the flow is typically assumed to be a turbulent flow over a rough surface; a surface that is sometimes assumed to be statistically homogeneous or, possibly, with an allowance for spatial inhomogeneities of the surface roughness. At the simplest level we can consider a statistically homogeneous rough surface. The fluid mechanics of this flow is generally taken to be well understood with an adiabatic flow adopting a logarithmic mean velocity profile over the lower section of the boundary layer but above the roughness sublayer within which the real in-homogeneities of the surface are felt as shown in Figure 2.4.

The logarithmic velocity profile scales on the friction velocity $u_*$ that, in turn, is determined by the surface stress. The effect of the surface roughness is felt through the friction velocity, the surface roughness length $z_0$, and a displacement length d. If the surface is non-adiabatic it is usually assumed that Monin-Obukhov similarity theory holds. Frequently the Monin-Obukhov length is not allowed to fall below some prescribed minimum value. This is a somewhat arbitrary way of treating nominally stable flow over a city but the heat release from the city at night will force the flow to be much less stable as will the larger roughness over the city. Tables of $z_0$ and, sometimes, d are available for particular land-use types (Hanna and Britter 2002). These may be used directly in simple operational models or indirectly through "wall functions" to provide boundary conditions for CFD/mesoscale models operating on the city or larger scales.

The flow near the urban surface may have a direction influenced by any order in the geometry of the surface and be different to the flow aloft.

Studies of the surface energy balance in urban areas (Grimmond and Oke 2002) lead to predictive techniques for the surface heat fluxes (in their various forms) and this is a required input for determining the Monin-Obukhov length. Other dynamically important parameters such as the surface temperature arise from the surface energy balance and these can be used with correlations for the occurrence of heat islands.

Dispersion of pollutants on the city scale involves pollutant plumes that are mainly deeper than the typical building heights and might therefore be treated in a manner that uses the gross flow parameters connected with the wind speed and direction and the turbulence levels and time scales (Hanna and Britter 2002). At an even simpler operational level there are correlations for dispersion coefficients that may be prescribed as functions of only "rural" or "urban" classes.

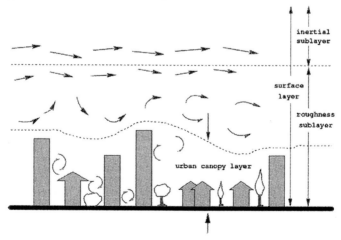

**Fig. 2.4.** The roughness sublayer (Britter and Hanna 2003)

This approach does not address the flow processes specifically within the roughness sublayer or within the urban canopy layer, and thus is unlikely to adequately describe the dispersion processes that occur within those layers. These dispersion processes are then those out to maybe 1 to 2 km; a scale which has sometimes been referred to as a neighbourhood scale. At this scale two approaches are apparent. The first is to use a parameterisation of the urban form appropriate for flow (Macdonald 2000) and dispersion while the second is to attempt computational studies that are obstacle-resolving (Schatzmann and Leitl 2002). Figure 2.5a is an aerial view of part of central London that could be considered to be a small neighbourhood. A digital elevation model of this area in Figure 2.5b shows the various building heights with a grey-scale. The street canyons are evident but the building height variation and the street intersections will complicate the idealised flow of Figure 2.3.

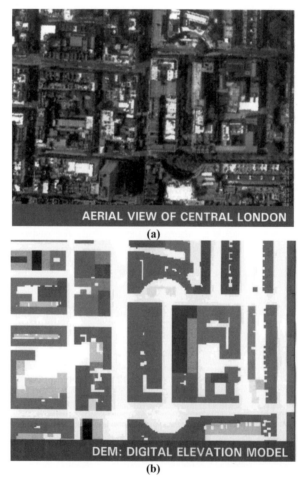

**Fig. 2.5. (a)** An aerial view of part of central London and **b)** a digital elevation model of the same "neighbourhood. With permission from Dr. C. Ratti

No studies within SATURN were directed towards the first approach. This approach has become very popular recently; the popularity stemming mainly through concern about the dispersion of accidental or deliberate releases of hazardous materials in urban or industrial areas. The second approach was used in several SATURN studies and can obviously be extended down to encompass the street scale considered in the previous chapter. Such computational studies then become a balance between the region to be studied, the level of geometrical detail required and the computing resources available.

An air mass may be over the city scale for many hours and over the regional scale for far longer. The relevant chemistry is considerably more complex with very many reactions and species. In particular the role of VOC's (volatile organic compounds) with $NO_x$ under photolysis in producing elevated ozone and $NO_2$ concentrations, that is photochemical smog (Moussiopoulos et al. 1995).

## 2.4 The interaction between the city and regional scales

The regional scale acts as the "background" for the flow on the city scale while the city scale acts to modify the regional scale flow. The city will divert the regional scale flow vertically and laterally both kinematically due to direct flow blockage and dynamically due to the increased drag force provided by the buildings. The city acts as a momentum sink for the regional flow and produces a modification to the velocity profiles. The city acts as a source of turbulent kinetic energy that will enhance vertical mixing over the city. The regional scale flow may also be dynamically affected through a thermally generated "heat island". The heat island produces a convergence into the city and vertical motions over the city. The regional scale provides the background chemical composition of the air within which the pollutants released in the city will mix, react and dilute. The city provides the pollutant source for an "urban plume" that can be detected for distances of 100 to 200 km downwind. Further chemistry changes the composition of the urban plume downwind.

Research in this area has several themes. One is the development of parameterisations of the surface processes for the city that can be used as boundary conditions for regional (mesoscale) models (Brown 2000), thereby restricting the need for extensive computer resources to be made available for the surface region. These boundary conditions may be in the form of drag coefficients or, more generally, surface exchange coefficients. Within the engineering CFD community similar coefficients are commonly referred to as "wall functions". The surface exchange processes are significantly different for momentum and, say, heat; the latter having no analogue with the forces arising on bluff bodies such as buildings due to the pressure differences across the building as a result of flow separation. Within this theme would also be the use of simple models for the flow within and near the city; a common example would be to model the city as a region of distributed porosity.

Another theme is the observation and study of particular combinations of the city scale processes and the regional scale flows that lead to severe pollution episodes. These may be of a generic nature or may be specific to particular cities. The effect of terrain can be particularly marked in producing local wind systems (Almbauer et al. 2000) including significant drainage flows.

However the main theme here is the development of regional (mesoscale) models (Schlünzen 2002) and their modification to accommodate the city-scale processes in general. Testing is carried out particularly with regard to ozone levels though other pollutants are also commonly used for testing. At the regional scale the chemistry is of great influence and much effort is expended on developing and testing accurate, robust and computationally acceptable chemical mechanisms.

## 2.5 Connecting the scales

The various physical and chemical processes are most easily discussed and analysed in isolation; the isolation being obtained above by addressing the processes of significance at each of the different scales. When considering urban air quality, particularly at the operational level, all the processes, from those at the smallest to those at the largest scales, may be significant. This said, professional judgement in balancing the required quality (fitness-for-purpose) of the solution approach to the problem and the required result frequently allows some of the scales to be omitted or very coarsely treated.

However most operational modelling and predictive studies requires the inclusion of all scales. Within SATURN studies were directed towards the inclusion of modelling at all scales. These projects, being essentially integration activities, did not lead to the specific study of physical or chemical processes but were obliged to consider the consistency of the modelling of the processes across the scales.

A related, important issue is how best to treat the boundary conditions at a particular scale. For example how should the input wind field be referenced for a street canyon study; in the canyon, atop local buildings, above the roughness sublayer, well above the roughness sublayer or the geostrophic velocity. Should it be the mean velocity or, possibly, the turbulence levels or a combination of several variables. Much the same could be said about studies at the neighbourhood or city scale.

Probably the most challenging requirement here is the inclusion of chemistry that is consistent across the scales. This may require modelling that adjusts itself to the scale naturally.

# Chapter 3: Air Pollutant Emissions in Cities

P. J. Sturm

Graz University of Technology, Institute for Internal Combustion Engines and Thermodynamics, Inffeldgasse 25, A-8010 Graz, Austria

## 3.1 Introduction

Emission inventories are important tools to describe the emission situation and eventually to manage air quality. They provide comprehensive information on emission sources and emission fluxes in the area under consideration. When looking at pollutant dispersion and air quality modelling it is necessary to have an accurate and high quality description of the emission sources in terms of quantity and dynamic behaviour. This requires an emission inventory with a high resolution in space and time. Within SATURN various applications of urban air quality models were planned. To ensure consistency among the emission inventories used for the validation attempts of the urban air quality models, guidelines with minimum requirements for emission inventories were defined in a joint activity between SATURN and the EUROTRAC-2 subproject GENEMIS-2 (Friedrich et al. 2003). These guidelines aim at (i) presenting a harmonised method for the compilation of urban emission data, (ii) providing sources for suitable emission factors, and (iii) providing tools for producing emission inventories. In this chapter a method is presented, which was developed for harmonising the procedure of setting up urban emission inventories (Sturm et al. 1998).

## 3.2 Requirements for urban emission inventories

The joint 'urban emissions working group' of the EUROTRAC-2 subprojects SATURN and GENEMIS-2 defined requirements that were considered as a minimum concerning quality of emission data and effort necessary to produce an appropriate emission inventory for urban air quality modelling. These requirements are briefly presented below. More details can be found elsewhere (Friedrich et al. 2003).

Pollutants: The atmospheric pollutants that need to be taken into consideration in assessment and management of ambient air are specified by Council Directives 96/62/EC of September $27^{th}$ 1996 and 99/30/EC of April $22^{nd}$ 1999 on ambient air quality assessment. Amongst these pollutants, $SO_2$, $NO_x$, CO, particulate matter (total and fine particulate matter), and hydrocarbons have to be considered as a minimum.

Emission sources: Usually, emission inventories classify emissions into three source categories, namely point, area, and line sources. In order to cover all activities which are related to emissions (regardless of the emission type), a nomenclature must be used. An appropriate framework that may be used for estimating emissions is the combination of SNAP 94 (Selected Nomenclature for Air Pollution), developed by EEA, and the EMEP/CORINAIR "Atmospheric Emission Inventory Guidebook" (EEA 1998).

When setting up an urban emission inventory, it is essential to define criteria for deciding whether an individual source is to be considered as a point source or to be combined with others to an artificial area source. This question is strongly linked to the methodology and models used to estimate quantitatively the emissions. Therefore, a clear separation has to be made between point and area sources, as well as between top-down and bottom-up approaches.

Emission factors: Data used to compile emission inventories can originate from a wide range of sources: Actual continuous measurements are usually available in the case of large point sources. Emission estimates may also be based on discontinuous measurements or on emission factors. For various activities, no direct measurements are available; therefore, so-called emission factors are being widely utilised when compiling emission inventories, especially in the case of emissions from traffic, small combustion, etc. In view of their decisive influence on the accuracy of the emission data, all emission factors used should always be clearly referenced.

Split of VOC and $NO_x$: Knowledge on the VOC subdivision and the reactivity of individual components is a prerequisite for reliable simulations of photochemical processes. Direct measurements of VOC species are too scarce and in any case insufficient for defining the VOC split in emission inventories. Therefore, VOC speciation is usually based on different relations established for various source categories. In order to make best use of state-of-the-art chemical mechanisms, a split into alkanes (methane, ethane, propane and higher alkanes), alkenes (ethene, terminal alkenes, internal alkenes, dienes and biogenic alkenes), aromatics (benzene, toluene and xylenes), and carbonyls (formaldehyde, acetaldehyde and higher aldehydes; ketones) is required.

Although mainly consisting of NO, nitrogen oxide emissions are given in emission inventories in terms of the equivalent $NO_2$. For the application of photochemical models it is necessary to specify the fractions of NO and $NO_2$ for each individual emission source.

Resolution in time and space: In principle, the temporal variation of the emissions follows the time dependence of the activity pattern. The standard for urban air pollution modelling is a time resolution of one hour, while a much higher resolution is necessary for microscale or local scale applications, as well as for resolving the situation at hot spots during air pollution episodes. Fig. 3.1 gives as an example the daily variation of VOC emissions from traffic on a certain winter day in Graz (Sturm et al. 1999). A lower temporal resolution may be sufficient for estimating the background concentration of primary pollutants. Towards this aim, it may be appropriate to distinguish between continuous and discontinuous sources, taking into account seasonal variations.

The spatial resolution is a characteristic feature of an urban emission inventory. As a general rule, the smaller the region of interest, the finer the required spatial resolution of the emission inventory. When focusing on local scale problems, the spatial resolution has to be as high as 250 m to 500 m, while for regional scale problems, typically tropospheric ozone simulations, a resolution of 1 km may be sufficient. Fig. 3.2 shows as an example $NO_x$ emissions from traffic during morning peak hour in Graz, adjusted to the numerical grid of an air pollution model (Sturm et al. 1999).

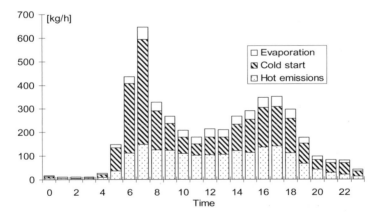

**Fig. 3.1.** Diurnal variation of VOC emissions from traffic in Graz on a typical winter day (Sturm et al. 1999)

**Fig. 3.2.** NO$_x$ emissions from traffic in Graz calculated for 8-9 LST* on 17 January 1995

# 3.3 Methodological approach

### 3.3.1 Urban scale emission inventories

Different methodologies can be applied to establish an emission inventory. Consideration of individual sources – based on *insitu* information on emissions and activity data – is referred to as "bottom-up" methodology. The use of statistical data, on the other hand, results in a "top-down" estimation.

The bottom-up methodology is based on a source-oriented inquiry of all activity and emission data needed for describing the emission behaviour of a single source. Top-down approaches are based on a disaggregation of existing emissions and/or activity data, and the use of statistical information from a coarser level. An exclusive use of the "bottom-up" method is in general impossible due to lack of available input data, while an exclusive use of the "top-down" method is not desired as it might lead to a poor level of accuracy of the urban emission inventory. For this reason, the two methods are usually combined.

For practical applications it is important to ensure that the bulk of the emissions are estimated in an accurate way (cf. Fig. 3.3): When considering industrial activities for the Antwerp area it can be seen that 20% of all point sources deliver 90% of the NO$_x$ emissions and almost all SO$_2$ emissions. If these 20% of point sources can be appropriately described the accuracy of the inventory will be sufficient. In order to achieve the level of accuracy required for urban air pollution modelling, it is recommended to use the methodological approach outlined in Table 3.1.

---

* LST: Local Standard Time

If the activity data is given on a statistical basis (e.g. for residential combustion), a top-down methodology must be applied. This is normally the case for SNAP sectors 2 (heat demand), 6 (solvents), 10 (agriculture) and 11 (nature). As a prerequisite for applying the bottom-up approach instead of the top-down method, sufficient information has to be available on location, activity data and emission factors. An example could be the bottom-up treatment of emissions from nature and agricultural sources when detailed information on land use cover and location is known and specific meteorological data and emission factors are available.

Furthermore, it is preferable to disaggregate emissions using land use information rather than administrative boundaries. For specified forest areas, emissions related to the surface area may be specified, and the same applies for traffic emissions not directly related to individual roads. Another example is that of emissions related to activities in settlement and built-up areas. Such emissions can be derived from statistical data.

Satellite data may be used for improving disaggregation procedures in emission inventories. A suitable method for this purpose was developed in the frame of the EU funded project IMPRESAREO (URL 3.1). Several test sites were chosen to optimise and to apply the spatial distribution of emissions following the approach developed within IMPRESAREO. The grid resolution used in these examples was consistently set to $1 \times 1$ km². The first applications referred to the cities of Stockholm and Linz (Austria), and were used to assess the weighting factors. These inventories were limited to just one pollutant ($NO_x$). Multi-pollutant inventories were created for Belfast and London, as well as for the whole England and Wales. The latter application proved that the method need not be restricted to urban areas. A final method evaluation, without further refinement, was performed for the Greater Athens Area and for the Milan province. For each of these areas, emission inventories are now available at different levels of detail, but at an identical spatial resolution.

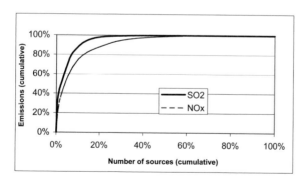

**Fig. 3.3.** Cumulative distribution of industrial point sources in the Antwerp area (Mensink et al. 1998)

**Table 3.1.** Methodology for treating emissions from different source types and activities

| Methodology | bottom-up | top-down |
|---|---|---|
| *Mobile sources* | | |
| Road traffic | Traffic on the main road network | Traffic on the secondary road network (area sources); cold start emissions and evaporation losses; activity data based on statistical information like fleet distribution per model year |
| Off-road traffic | | Disaggregation based on statistical data |
| Railway traffic | Regular traffic on the rail network | Shunting activities |
| Maritime and inland traffic on waterways | Scheduled traffic (ferry boats etc.) | Harbour activities |
| Aviation | Methodology in analogy to rail and water | |
| *Stationary sources* (according to selection criteria) | | |
| | Point sources | Area sources |

### 3.3.2 Local scale emission modelling

Local scale applications require a quite different methodology as they aim to estimate emissions at a much higher detail level. Basically, there are two different purposes of that type of models. One is based on the requirements of traffic planners, who want to simulate traffic flow and fuel consumption/emissions on a vehicle-by-vehicle basis. Such models are often used for investigations of traffic flows between crossings, traffic signals, etc. A second application type is related to air quality simulations, where local scale dispersion models need an appropriate emission input at the same scale. Ideally, one single microscale traffic/emission model could provide the input for both applications. Unfortunately, the currently available emission data is not accurate enough to deliver such detailed emission estimates at the accuracy needed. There are several reasons for this shortcoming:

- Description of the actual fleet: Fleet data is statistical in nature, i.e. observations, models and/or counts deliver only a rough split between different vehicle categories, but almost no information about emission standard, engine technology, etc. Yet, such data would be needed for an accurate fleet description.
- The working conditions (especially the temperature) of the engines are not known. In real world, most in-city journeys are operated under conditions with cold or warming-up engines. Correspondingly, neither the engine, nor the engine/exhaust gas after treatment system work properly. The share of vehicles operating under such conditions is not known and, therefore, the cold surplus emissions cannot be allocated to the right street parts.
- Microscale emission models require instantaneous emission data. Such data is available for pre EURO-2 vehicles, but their application is limited (Sturm et al. 1998). New technology vehicles with very low emission levels are extremely difficult to model on an instantaneous second-by-second basis.

## 3.4 Quality assurance

Quality assurance should be by all means accounted for when setting-up an urban emission inventory, and especially in case that such an inventory is supposed to be used for urban air quality modelling. Details on this topic are given by Cirillo et al. (1996). The remainder of this section corresponds to a short summary from the above reference, extended and adjusted for urban scale applications.

All information compiled under the heading "quality control/quality assurance" needs to be made available at least within the respective project in order to warrant the conditions of its collection. A quality management person must be appointed within each project to be responsible for assessing to what extent established guidelines have been fulfilled.

A completeness analysis shall ensure that all activities - anthropogenic and natural - which generate emissions of a certain pollutant have been considered, or at least the emission sources not considered are negligible with respect to the considered time and spatial scales.

Uncertainty assessment is essential for any quality evaluation. This is related to aspects such as the indicator choice, the quantitative value of the indicators, the emission factors and the representativeness of the emission factor for the particular sources to which it is applied.

Validation of an urban emission inventory should be attempted with the help of alternative emission assessments including analysis of emission trends vs. ambient air concentration trends, flux estimations from downwind measurements and application of atmospheric models.

Documentation is a main issue in context with quality assurance. The documentation has to cover the methods used, emission factors, sources for activity data and notes on the quality assurance procedure itself.

## 3.5 Uncertainty

In general, uncertainty assessment is part of the quality assurance procedure (see above). Emission inventories are always based on mathematical assumptions and statistical data. Real measurements are available for a few emissions sources and/or for certain time periods only. Therefore, uncertainty estimations are of importance and should always be foreseen. However, it is not easy to assess uncertainty at the level of aggregated datasets. Information related to emissions and their uncertainties that originates from a few measurements has to be applied for a large number of sources. As the latter may differ strongly, even within the same category, the uncertainty increases as the emission inventory becomes more detailed. On a more aggregated level, averaging helps to improve the uncertainty situation. This is shown in Fig. 3.4 where the uncertainty of emission estimates for passenger cars is given. If someone has to consider the emission situation on a certain street, then two problems have to be tackled. First of all it is not known which individual vehicles pass this street section. In the rare case that this information is available the respective emission data will not be available, as this data comes from measurements of similar but other vehicles and these may have different emission characteristics due to various reasons such as different maintenance standards and varying driving styles. The second point is that the driving behaviour and hence the emission situation may vary for each vehicle (left hand part of Fig. 3.4). Measurements of emissions from the same vehicle passing the same street at different times of the day show that the emissions (in this case $NO_x$) vary between –90% and +260%. In other words, the emissions can either be a tenth of the average or almost three times higher. As soon as a first averaging is made to cover the average fleet vehicle in an average street of certain characteristics, the uncertainty decreases to a range between –20% and +30%. More or less the same uncertainty exists if total passenger car traffic inside a city is considered. An assessment of uncertainty of emissions from road traffic as a whole, both on motorways and in urban areas, came up with more or less the same results (Fig. 3.5; Kühlwein and Friedrich 1999).

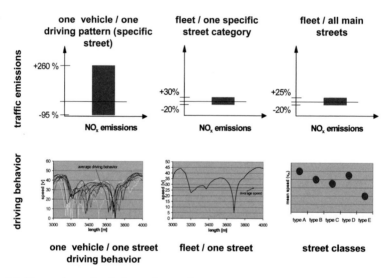

**Fig. 3.4.** Uncertainty of emission estimates for passenger cars as a function of the aggregation level (Sturm 2000)

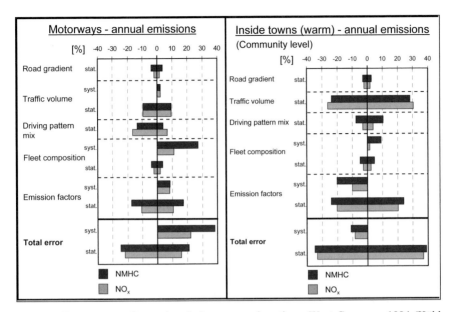

**Fig. 3.5.** Error ranges of annual emissions on road sections, West Germany, 1994 (Kühlwein and Friedrich 1999)

An uncertainty assessment was performed for the stationary sources included in the emissions inventory of Graz. This assessment came up with the uncertainty ranges given in Table 3.2. The closer the emission factor can be related to the source type (e.g. through measurements at a single source), the better the quality. The opposite situation appears if the emission estimate has to be based on emission factors related to the generic activity type rather than the specific situation.

**Table 3.2.** Uncertainty ranges for stationary sources (Sturm et al. 1999)

| Source type | Methodology | Uncertainty range |
|---|---|---|
| Combustion related emissions | using measurement data | +/- 35% |
| | using fuel consumption (and type) | +/- 40 % |
| | using thermal fuel efficiency | +/- 65% |
| | using energy consumption figures based on number of employees or material throughput | +200% - 90% |
| Process related emissions | using measurement data | +/- 35% |
| | using emission factors based on industrial type, number of employees or material throughput | +/- 40% |

## 3.6 Applications

This section includes three examples for emission inventories designed for providing input to dispersion models.

### 3.6.1 Emission inventory of the Antwerp area

The urbanised triangle Antwerp-Brussels-Ghent is located in Northwest Europe and can be characterised by open, flat terrain conditions. The emission inventory for the City of Antwerp is largely influenced by its harbour with a relatively large amount of refineries and chemical industry. Besides the contribution of industrial emissions, the ring road of Antwerp generates a substantial contribution from traffic emissions, due to the fact that this highway is part of the transit route from the Netherlands to France.

An urban emission inventory (CO, $NO_x$, VOC, PM, $SO_2$, $CO_2$, $N_2O$, $CH_4$, $NH_3$ and heavy metals) has been developed for the Antwerp area, which includes industrial sources, road transport and space heating. The spatial resolution is determined by the national emission inventory, as provided by the Flemish Environmental Agency (VMM 2001), containing yearly averaged emission data for point sources, line sources, (motorways and national main roads) and $1 \times 1$ km$^2$ area sources for space heating and urban road transport. Rather than using the national traffic emission inventory, a road transport emission model was applied, which was specifically designed for the City of Antwerp (Mensink 2000). The temporal emission variations are modelled using temporal emission factors based on various sources of emission measurements and national statistics (e.g. traffic counts and several years of degree-days measured at different locations). The inventory has been validated partially by a comparison with emission measurements (Mensink 2000). As an example, Fig. 3.6 shows calculated benzene emissions during one week in March 1988.

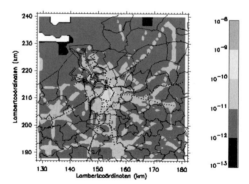

**Fig. 3.6.** Benzene emission fluxes (kg m$^{-2}$ s$^{-1}$) in the Antwerp area during 23-27/3/98

## Spatial emission distribution

### Industrial point sources

The emission inventory for the Antwerp area can be characterised by the presence of a relatively large contribution of industrial point sources. In 1996, 72% of the $NO_x$ ($NO_2$) emissions and 94% of the $SO_2$ emissions in the urban area of Antwerp could be allocated to industrial point sources. Since 1993, industrial companies are obliged to report their emission yearly to the Flemish Environmental Agency. Based on these registrations and on additional calculations and estimates using statistical data and emission factors from the literature, the Flemish Environmental Agency composes a yearly emission inventory of industrial point sources.

For the Antwerp area ($20 \times 20$ km²), in 1996 208 point sources were registered to emit $NO_x$ ($NO_2$) and 174 to emit $SO_2$. The cumulative distribution of the industrial point source emissions, both for $NO_x$ and for $SO_2$ has already been discussed in conjunction with Fig. 3.3. This figure reveals that only a small fraction (15-30%) of the industrial point sources is responsible for 95% of the $SO_2$ and $NO_x$ ($NO_2$) emissions in the urban area of Antwerp.

### Space heating

In contrast to the individual registration of industrial point sources, the emissions due to space heating are evaluated by a collective registration per km². The inventory consists of area sources for private households and non-private buildings and is determined by a top-down approach. The emissions for private households are based on a 10 yearly census of houses per statistical unit. A statistical unit contains more or less the same building types. Furthermore, the emissions are calculated depending on the type of building (e.g. apartment, family residence), the type of heating system and type of fuel used. Non-private buildings are split into collective housing (e.g. hospitals), schools, public buildings (e.g. offices), shops and greenhouses.

### Transport emissions

Much attention is paid to road transport emissions. They are derived by a "bottom-up" approach using VITO's on-the-road traffic emission measurements (De Vlieger 1997) in combination with an urban traffic flow model, which is actually used by the Antwerp City authorities and includes statistics on traffic numbers and vehicle types for a network of almost 2000 road segments. The emission factors are partially derived from the on-the-road traffic emission measurements as well as from COPERT-III (Ntziachristos and Samaras 2000). Distinction was made between hot emissions, cold emissions and emissions due to evaporation (diurnal losses, running losses etc.). Five classes of vehicles are distinguished (passenger cars, Light Duty Vehicles, Heavy Duty Vehicles, motorcycles and mopeds) with a further subdivision depending on the age of the vehicle and its cylinder capacity for the passenger cars and for the two-wheels vehicles, while age and weight criteria are used to distinguish LDV and HDV (including busses and coaches). The operated classification is analogous to the sub-classification proposed in the COPERT-III methodology, however limited to 105 classes in total. Distinction is made between four fuel types (gasoline, diesel, LPG and 2-stroke gasoline), with lead and sulphur contents depending on the simulation year.

The road network is divided in six road classes (highways, national roads, main roads outside the city, main roads inside the city, secondary roads and harbour area). A specific year-dependent fleet composition is associated to each road class. The relative fleet composition has been updated for the years 1996 to 2000 using the most representative and actual data. Estimated vehicle park compositions for the years 2005 and 2010 were also implemented. The road network data-file consists of the Lambert coordinates of each road segment, together with the number of vehicles on this segment and their hourly average speed. The average trip length used for the calculation of the cold emissions is automatically determined from the data-file.

Maritime traffic emissions are not included in the emission inventory for the Antwerp area. Comparisons of calculated $SO_2$ concentrations with measurements, however, show that the contribution of these emissions is not negligible. As yet the information available is insufficient to estimate the emissions associated with harbour activities in Antwerp.

### *Temporal emission distribution*

Emission time factors $\Gamma_m$ (monthly variation), $\Gamma_d$ (daily variation) and $\Gamma_h$ (hourly variation) are determined per emission category, such that:

$$(12 \cdot 7 \cdot 24)^{-1} \sum_{m,d,h} (\Gamma_m \cdot \Gamma_d \cdot \Gamma_h) = 1$$

Time factors for point sources were derived from the Emission Handbook (Mc Innes 1996). Daily variations in space heating were not investigated. Monthly variations were based on calculated degree days for Antwerp. The time factors for road transport were derived from traffic counts on 18 locations.

### 3.6.2 Emission inventory of the city of Graz

In this section the emission inventory of the city of Graz is briefly presented. A detailed description of this inventory is given by Sturm et al. (1999).

The first emission inventory for Graz was created in 1989. Apart from the main objective of quantifying emissions, this inventory also served as an input for an urban air quality model. A precondition for this type of purpose is a high temporal and spatial resolution, in this case, one hour and 250×250 m, respectively. In addition, it is necessary to define emission sources at a very detailed level in terms of pollutants, source strengths, activities and locations. Updates of the emission inventory were provided for the base years 1995 and 1998 for two main reasons. Firstly, to make a new estimation of the emission quantities and in this way to define the trend for previous years and secondly, to have an up-to-date emission inventory for on-going research projects dealing with pollution dispersion in the city of Graz.

Details on the traffic related part were already discussed in conjunction with Fig. 3.1 and Fig. 3.2. Traffic plays a major role regarding $NO_x$ emissions, while the private sector (households and small business) is responsible for the major part in energy consumption (and hence $CO_2$), $SO_2$ and CO emissions. The relative share of the sectors traffic, industry, and private (small combustion processes) is characteristic for each city (Fig. 3.7). The ecological and climatological conditions of the city of Graz are responsible for this specific emission split.

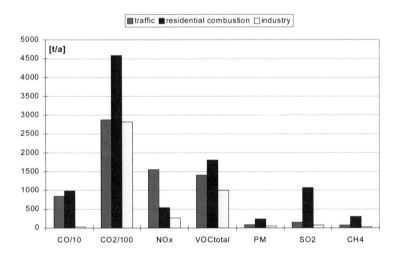

**Fig. 3.7.** Emission share of the individual source types for the city of Graz

In order to analyse emission trends it is important to update emission inventories. An update of an emission inventory is always based on a better knowledge of emission and activity data, but also on improved emission models. A misinterpretation should be avoided by using the same emission models and emission factors (provided that the relevant emission technology is taken into account) for the different base years. In other words, to identify the trend between two base years it is necessary to recalculate the old situation with the models used for the new one. Fig. 3.8 shows the potential discrepancy if an emission inventory is updated and a change in the databases is accompanied by changed models. For that specific year

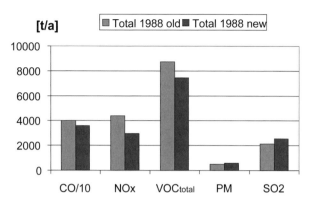

**Fig. 3.8.** Differences in emission quantities due to changes in emission/activity data bases and models

(1988) the differences are not that large (some 5 to 20%), but only because the changes between the source types balanced out. The traffic sector, for instance, experienced remarkable $NO_x$ reductions, while the small combustion sector had increases of almost the same magnitude. So in total the difference was small but for the different source types and pollutants the difference due to technological change and variation in the usage patterns was up to ±50%.

### 3.6.3 Emission inventory for the Lombardia Region

The Lombardia Region is located in Northern Italy and includes several cities (examples: Milano, Bergamo, Brescia, Varese, Como) as well as rural areas in the Po Valley. The domain is densely inhabited and includes highly industrialized and intensive agricultural areas.

This area is regularly affected by high ozone levels, due also to frequent air stagnation conditions, especially during summer months. This fact requires a careful design of emission abatement strategies, taking into account photochemical regimes, emission and meteorological conditions. Therefore, a reliable Lombardia area emission inventory is fundamental to provide emission fields as input to photochemical modelling system simulations; an emission model is needed to evaluate the spatial sources distribution, the modulation in time, the NMVOC splitting and lumping, the PM size distribution and chemical description according to photochemical model requirements.

The emission processor POEM-PM (Catenacci et al. 2000; Finzi et al. 2002) has been specifically designed to produce present and alternative emission fields estimated by means of an integrated "top-down" and "bottom-up" approach. POEM-PM can be interfaced in particular with the Italian CORINAIR data and takes into account diffuse and main point sources coming from different activity sectors. Model outputs are the results of all possible combinations of four steps: spatial disaggregation, time modulation, NMVOC and PM splitting. As far as spatial disaggregation is concerned, the model estimates municipality emissions starting from the province level. It is also possible to distribute emissions on a grid domain; it is so allowed to design future scenarios adding new sources configuration, deleting point sources, introducing new roads or pedestrian areas and changing road network. The spatial allocation makes use of *surrogate variables*, highly correlated with emissions and defined by means of national and local statistical sources, GIS and land-use information. Regarding time modulation, the model can provide emission fields for any assigned time interval, on a daily or hourly basis, starting from the annual database. In accordance with the EUROTRAC-2/GENEMIS-2 subproject, fuel use, temperature, degree-days, working time, production cycle, traffic counts and road statistics are the main indicators used for the temporal modulation of emission activities. The total VOC amount is split into SAROAD classification of individual compounds and then lumped into emission classes according to the photochemical mechanism implemented in the transport model. The FCM interface has also been implemented in POEM-PM model to allow flexibility in chemical schemes. Using suitable speciation profile, total PM emissions are split into 10 dimensional classes, from 0 to 11.39 µm, and 6 chemical categories (Organic Carbon, Elementary Carbon, Sulphates, Nitrates, Water, Other). Because epidemiological studies have shown that fine particles (diameter < 1 µm) have a deep impact on human health, the size classification is focused on small diameter classes.

PM$_{10}$ emission field have been estimated for a working summer day (1$^{st}$ June 1998). Simulations show considerable emissions between 6 am and 12 pm and a peak at 7 pm, when traffic is more intensive (see Fig. 3.9); moreover emissions are concentrated near urban areas (mainly Milan metropolitan area) and main extra urban roads.

The emission model has been validated on simulations of the air quality impact by means of GAMES (Gas Aerosol Modelling Evaluation system) (Decanini et al. 2002). The episode chosen was the period from 1 to 5 June 1998 (Gabusi et al. 2001), occurred during the PIPAPO projects measurements campaign performed in the frame of EUROTRAC-2. During the selected period, values exceeding the health protection threshold have been recorded in many stations of the regional network. The comparison with monitored data has revealed the capability of the whole system to reproduce the main spatial and temporal features of O$_3$ and PM distributions.

As a consequence of its technology and fuel-oriented formulation, the emission model POEM-PM can be used to provide scenarios consistent with new fuel trades and pollutant abatement technologies. In particular, some pollutant control technologies (catalytic converter, zero-emission vehicle) and alternative clean fuels (oxygenated fuels, compressed natural gas), together with their specific splitting profiles, can be taken into account for sustainable air quality improvement strategies. Several projects have been devoted to estimate the emissions related to different traffic policies. For example the impact of EU Directives on road traffic emissions scheduled for 2005 has been investigated (Volta et al. 1999); the study results in terms of emission rate reductions are reported in the Table 3.3. By taking into account the distribution of these emission sectors over the domain (including area and point sources), the total reduction is estimated at about 25% for NO$_x$ and 15% for NMVOC and CO.

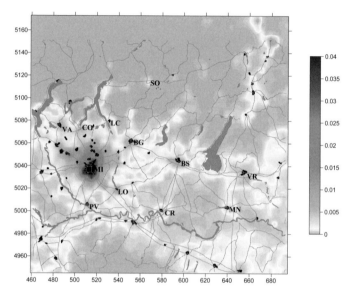

**Fig. 3.9.** PM emission (t/h) field estimated in Lombardia region for 1$^{st}$ June 1998 (7.00 p.m.)

**Table 3.3.** Emission reductions in the Lombardia region between the 2005 and the base-case (1996) scenarios

|             | NOx      | CO       | VOC      |
|-------------|----------|----------|----------|
| highway     | -55.23%  | -47.73%  | -53.34%  |
| extra-urban | -49.76%  | -47.81%  | -45.25%  |
| urban       | -30.48%  | -46.21%  | -37.20%  |

## 3.7 Intercomparison of urban emission inventories

Various urban emissions inventories were set up in the framework of the SATURN and GENEMIS-2 subprojects. Amongst others, these inventories should also provide input to dispersion models. In particular, inventories covered the cities/regions indicated in Table 3.4.

**Table 3.4.** Urban emission inventories set up in EUROTRAC-2

| City | Prepared by |
|------|-------------|
| Antwerp (B) | VITO |
| Athens (GR) | Aristotle University Thessaloniki |
| Augsburg (D) | University Stuttgart |
| Barcelona (E) | Politechnica by Catalunia |
| Graz (A) | "Erzherzog Johann" Graz University of Technology |
| Linz (A) | Austrian Research Centre Seibersdorf |
| Lisbon (P) | University of Aveiro |
| Milano-Lombardia (I) | University of Brescia |

**Table 3.5.** Intercomparison of urban emission inventories; pollutants treated

| City | Antwerp | Athens | Augsburg | Graz | Linz | Lisbon | Milano |
|------|:-------:|:------:|:--------:|:----:|:----:|:------:|:------:|
| CO2 | | | | ✓ | ✓ | ✓ | ✓ |
| CO | ✓ | ✓ | ✓ | ✓ | ✓ | ✓ | ✓ |
| NOx as NO2 | ✓ | | ✓ | ✓ | ✓ | ✓ | ✓ |
| NO | | ✓ | | | | | ✓ |
| NO2 | | ✓ | | | | | ✓ |
| SO2 | ✓ | ✓ | | ✓ | ✓ | ✓ | ✓ |
| HC | | ✓ | | | | | |
| NMHC | | ✓ | | | | | |
| VOC | | | | | | | |
| NMVOC | ✓ | ✓ | ✓ | ✓ | ✓ | ✓ | ✓ |
| CH4 | | ✓ | ✓ | ✓ | | | ✓ |
| VOC specification | | ✓ | ✓ | ✓ | | ✓ | ✓ |
| Particles (TSP) | ✓ | | | ✓ | ✓ | | |
| Particles (PM10, PM2.5, Black Carbon) | | | | | | | |
| heavy metals | ✓ | | | | | | |
| POPs | | | | | | | |
| N2O | | | | | * | | ✓ |
| NH3 | | | | | * | | ✓ |

**Table 3.6.** Mobile emission sources covered in the different cities

| | Antwerp | Athens | Augsburg | Graz | Linz | Lisbon | Milano |
|---|---|---|---|---|---|---|---|
| Road traffic | ✓ | ✓ | ✓ | ✓ | ✓ | ✓ | ✓ |
| Off-road traffic | ✓ | | ✓ | | ✓ | | |
| Rail traffic | | | | ✓ | ✓ | | |
| Aiviation | | ✓ | | | ✓ | | ✓ |
| Waterborn traffic | | ✓ | | | ✓ | | |

This variety of inventories offered the possibility to check if comparability between emission inventories exists in terms of emissions per activity, population or other socio-economic data.

The first step in the intercomparison tackled the topic of completeness, i.e. sources treated and pollutants covered. Already at this stage remarkable differences were identified, as each city/region has its own emission characteristics and therefore the priorities were set on different topics.

Table 3.5 contains as an example a template with some general information about pollutants that were included in the inventories. As it can be seen from the table some cities deal with $CO_2$, all of them include information about hydrocarbons but some of them state it as HC whilst others state it as VOC, although there can be remarkable differences, e.g. when considering HC emission factors from diesel cars.

Table 3.6 gives some information about emission sources covered in the different cities. For each of the cells additional information is available revealing how the emissions were calculated and which activities are covered – e.g. hot emissions, cold start, evaporation losses.

If qualitative assessments are made it becomes already difficult to perform comparisons. As was noticed in conjunction with Tables 3.5 and 3.6, not all inventories cover the same pollution sources. Specifically, when considering traffic, in some cities all traffic modes are included, while in others mostly road traffic is concerned and the others neglected. This is of course again a question of necessity. A city like Antwerp has huge harbour activities and has to treat the emissions from this source, while others do not. Another difference, which should be mentioned, is that not all emission inventories were based on the same year. This may become a problem when a year is used which is rather untypical in terms of e.g. domestic heating, as a city's individual characteristic will not be reflected adequately.

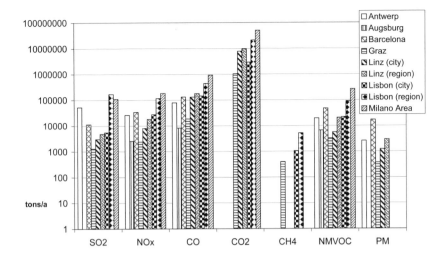

**Fig. 3.10.** Comparison of total emission amounts for different cities/regions

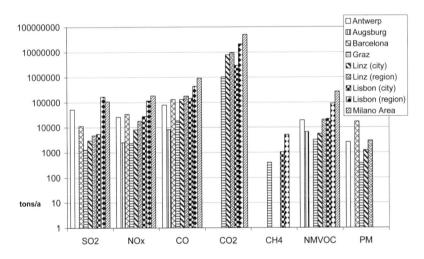

**Fig. 3.11.** Comparison of emissions from industrial activities for different cities/regions

Fig. 3.10 illustrates the total emissions for several cities. This shows already the variety of sizes as small cities like Graz and Linz were merged with large areas like Antwerp or Milano. This is of course reflected in the emission quantities. (It should be noted that the scale in Fig. 3.10 is logarithmic.) If the total emissions are broken down to normalised figures on a basis of per capita or per area, the results reflect the unique emission patterns of these regions. This can be seen from Fig. 3.11 where the emissions from the aggregated sector industrial activities are depicted. Cities/regions like Milano, Linz or Antwerp have high industrial activities and hence a high-energy consumption and emission levels. This is even more pronounced when looking at the share of emissions from industrial activities (Fig. 3.12). All these differences in activity patterns of the various regions are clearly reflected in the emission situation of the cities/regions.

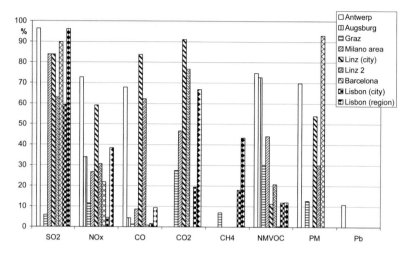

**Fig. 3.12.** Share of emissions from industrial activities for different cities/regions

## 3.8 Summary

When considering air pollution in cities two questions have to be tackled. One is the emission situation and the second is air quality. Both are linked with dispersion (transmission) processes and chemical reactions. Starting point of each air quality study for an urban area is the inventory of the emission sources. If emission information is to be used for dispersion modelling and air quality studies, it is necessary for such a specific emission inventory to fulfil certain requirements. These requirements cover the emission source strength as well as the spatial allocation and temporal variation of the sources. Quality assurance procedures and uncertainty assessments have also to be accounted for. Minimum requirements have been set up within the EUROTRAC 2 subprojects SATURN and GENEMIS-2 and the methodology developed was utilised to set out emission inventories for several cities/regions. These inventories were analysed and compared in order to get insight to the emission situation in various cities in Europe.

# Chapter 4: Urban Field Campaigns

P. Mestayer[1], R. Almbauer[2], O. Tchepel[3]

[1] Equipe Dynamique de l' Atmosphere Habitee, Laboratoire de Mecanique des Fluides, Ecole Centrale de Nantes,B.P. 92101, F-44321 Nantes Cedex 3, France

[2] Graz University of Technology Institute for Internal Combustion Engines and Thermo-dynamics, Inffeldgasse 25, A-8010 Graz, Austria

[3] Department of Environment and Planning, University of Aveiro, P-3810 Aveiro, Portugal

## 4.1 Introduction

One of the primary aims of the field campaigns within SATURN was to provide measuring data of good quality allowing evaluation of numerical tools. Dispersion models need to be validated in various urban environments and for the whole range of meteorological conditions occurring in the real atmosphere. However, experimental data for model validation are particularly scarce for some specific conditions such as a stable atmosphere with calm or light winds, coastal breeze and mountain-valley circulation, etc. In order to enrich existent set of validation data, experimental campaigns within SATURN have been designed. Related activities contributed also to a significant improvement of our understanding of airflow, dispersion and chemical transformation processes at the local and urban scales.

This chapter summarises the results of field experimental activities planned and executed in the frame of SATURN. The experiments are subdivided into local scale campaigns, urban scale campaigns and urban experimental activities of monitoring character. The attempt is made that for all campaigns the major objectives of measurements are identified, the experimental set-up completely described and the set of data obtained properly presented and discussed.

## 4.2 Local scale (street scale) campaigns

### 4.2.1 Runeberg St., Helsinki

The experimental campaign in the street canyon Runeberg Str. in Helsinki developed a dispersion dataset gathered during the intensive measuring period and is suitable for evaluation of street scale models. The street canyon dispersion model OSPM was evaluated against this dataset in co-operation of Finnish Meteorological Institute (FMI), Helsinki Metropolitan Area Council (YTV) and National Environmental Research Institute, Denmark (NERI).

### *Major scientific issues investigated (Runeberg Str., Helsinki)*

In climatic conditions of Northern Europe a stable atmospheric stratification with light wind speeds may prevail for extensive periods. For instance, it is not uncommon for such conditions to last for several days in southern parts of Finland, particularly in winter and spring (Kukkonen et al. 1999; Karppinen et al. 2001). The main objective of this work was thus to produce good-quality street canyon data, which could be utilised for model evaluation especially in light wind speed conditions.

The street canyon dispersion model OSPM was evaluated against this dataset for the so-called intensive measurements period (Kukkonen et al. 2000 and 2001a). A preliminary discussion of the measurements during the entire experimental period (one year) is presented elsewhere (Wallenius et al., 2001). The data has also been used for evaluation of the combined application of microscale traffic simulation and street canyon dispersion models (Granberg et al. 2000).

### *Relevance to SATURN aims (Runeberg St., Helsinki)*

The measurement campaign provides street canyon data that can be utilised for model evaluation. The dataset can also be utilised for studying the chemical transformation processes of $NO_x$ and $O_3$.

### *Experimental set-up (Runeberg St., Helsinki)*

In 1997, a measuring campaign was conducted in a street canyon (Runeberg St.) in Helsinki. Hourly mean concentrations of CO, $NO_X$, $NO_2$ and $O_3$ were measured at street and roof levels, the latter in order to determine the urban background concentrations. The relevant hourly meteorological parameters were measured at roof level; these included wind speed and direction, temperature and solar radiation. Hourly street level measurements and on-site electronic traffic counts were conducted throughout the whole of 1997. Roof level measurements were conducted for approximately two months, from 3 March to 30 April in 1997, the so-called intensive measurements period. The experimental set-up is illustrated in Fig 4.1.

### Main results (Runeberg St., Helsinki)

The OSPM model was used to calculate the street concentrations and the results were compared with the measurements. The overall agreement between measured and predicted concentrations was good for CO and $NO_x$ (fractional bias were -4.2 and +4.5 %, respectively), but the model overpredicted the measured $NO_2$ concentrations (fractional bias was +22 %).

The agreement between the measured and predicted values was also analysed in terms of its dependence on wind speed and direction; the latter analysis was performed separately for two categories of wind velocity. An example of these analyses is presented in Fig. 4.2a and b, in which the measured and modelled normalised $NO_x$ concentrations are shown, plotted against the wind direction. The background concentrations measured at the roof station have been subtracted from the street level concentrations, and this concentration has been normalised by the actual emission value, (C(street level) – C(roof level))/Q. The data presented are from daytime hours only (from 6 to 23 hours).

The differences between normalised measured and predicted concentrations can be attributed to the uncertainties in the concentration measurements, to the uncertainties in the estimations of the emissions and to the inaccuracies of the dispersion model.

Fig. 4.2 shows a clear dependency of both measured and predicted concentrations on the wind direction. This dependence is much more pronounced for higher wind speeds (u > 2 m $s^{-1}$) than for lower wind speeds (u ≤ 2 m $s^{-1}$). When the measuring point is on the windward side, concentrations are substantially lower, compared with the corresponding results for the leeward side.

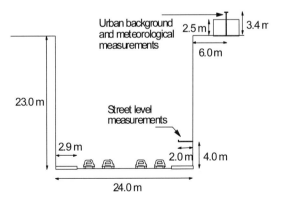

**Fig. 4.1.** Vertical cross-section of the Runeberg street canyon in Helsinki showing the locations of the measurement points at street and roof levels

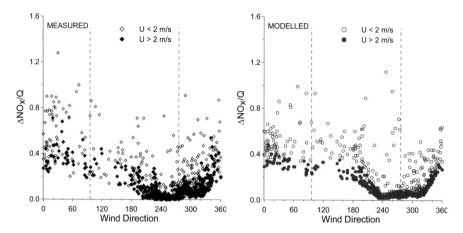

**Fig. 4.2.** Influence of the wind direction on the measured normalised hourly $NO_x$ concentrations, (C(street level) – C(roof level))/Q, where Q is the emission strength, and corresponding predicted normalised $NO_x$ concentrations. The dotted vertical lines indicate wind direction perpendicular to the street

Although measurements are only available from one side of the street, the clear difference between the leeward and windward sectors provides strong evidence for the formation in the street of a vortex in conditions of higher wind speeds. The flow structure in the street definitely might be more complex than the simple assumption of a single vortex made in the OSPM model, but the experimental data presented here does not provide any clear evidence against this simplification.

In the case of low wind speeds, there is no clear dependence of concentrations on wind direction, according to both measured and predicted results. Evidently, the formation of a stable vortex is much less likely at lower wind speeds, and this results in a more homogeneous distribution of pollution across the street canyon.

The database, which contains all measured and predicted data, is available for further testing of other street canyon dispersion models. The dataset clearly contains a larger fraction of low wind speed cases, compared with datasets from previous street canyon measurement campaigns conducted, for instance, in Denmark. A limitation of this measurement campaign is that no flow or turbulence measurements within the street canyon are available. The concentration measurements also do not contain particulate matter.

## 4.2.2 Elimäki, Finland

The experimental campaign in a roadside environment in Elimäki in Southern Finland developed a dispersion dataset suitable for the evaluation of street scale dispersion models. Graz University of Technology, Austria, and Finnish Meteorological Institute (FMI) have evaluated a Gaussian finite line source dispersion model (CAR-FMI) and a Lagrangian dispersion model (GRAL) against this data set (Öttl et al. 2001).

### *Major scientific issues investigated (Elimäki, Finland)*

The experimental campaign was designed specifically for model evaluation purposes. In this campaign, the dispersion of NO, $NO_2$ and $O_3$ was investigated on both sides of a road depending on traffic densities and relevant meteorological parameters (Kukkonen et al. 2001b).

### *Relevance to SATURN aims (Elimäki, Finland)*

The measuring campaign provides roadside data that can be utilised for model evaluation. The dataset can also be utilised for studying the chemical transformation processes of $NO_x$ and $O_3$. The dataset is properly documented and it is available for evaluation of other roadside dispersion models.

### *Experimental set-up (Elimäki, Finland)*

The experimental set-up is illustrated in Fig. 4.3.

### *Main results (Elimäki, Finland)*

The agreement of measurements and CAR-FMI model predictions was good, taking into consideration various statistical parameters. For all data (N = 587), the index of agreement (IA) was 0.83, 0.82 and 0.89 for the measurements of $NO_x$, $NO_2$ and $O_3$, respectively.

The difference between model predictions and measured data in terms of meteorological parameters was also investigated by Kukkonen et al (2001b). At wind speeds lower than approximately 2 m s$^{-1}$, there are excessively high predicted concentrations, compared with the measured data. Model performance deteriorates for situations with the lowest wind speeds.

The model overprediction at low wind speeds could be caused by the assumption of steady-state, homogeneous wind flow made in Gaussian line source models. The variation in wind direction tends to increase with decreasing wind speed, causing increased plume meandering. Substantial meandering can cause measured concentrations to be much lower, compared with those computed assuming a homogeneous wind flow (e.g., Benson 1992).

Öttl et al. (2001) extended the above mentioned evaluation, by comparing the performance of the Lagrangian dispersion model GRAL against the Elimäki dataset; this yielded information on the validity of two types of dispersion models. The agreement of measured and predicted datasets was good for both models considered, according to various statistical parameters used. For instance, considering all $NO_x$ data, the index of agreement values varied from 0.76 to 0.87 and from 0.81 to 1.00 for the CAR-FMI and the Lagrangian models, respectively.

The above mentioned Lagrangian model provides special treatment to account for enhanced horizontal dispersion in low wind speed conditions; while such adjustments have not been included in the CAR-FMI model. This type of Lagrangian model therefore predicts lower concentrations in conditions of low wind speeds and stable stratification, in comparison with a standard Lagrangian model.

Öttl et al. (2001) also analysed the difference between the model predictions and measured data in terms of the wind speed and direction. Fig. 4.4 shows data gathered in one specific wind direction quadrant, extending from a wind direction perpendicular to the road (air flow from a direction of 120°) to a wind direction parallel with the road (air flow from a direction of 210°, cf. Fig. 4.3). For simplicity, wind direction has been normalised, with a direction of 0° corresponding to the direction parallel to the road and 90° to that perpendicular to the road.

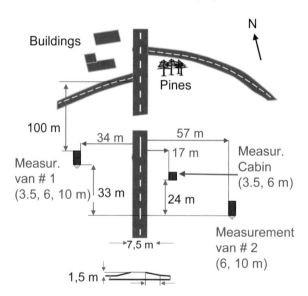

**Fig. 4.3.** Scheme of the measuring units and their environment in the roadside campaign at Elimäki, Finland. The measuring heights have also been indicated in parentheses

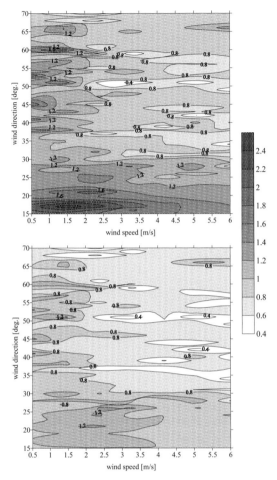

**Fig. 4.4.** The dependency of the ratio of predicted and observed concentrations on wind speed and direction for the CAR-FMI (top) and GRAL (bottom) models for the measurement site at Elimäki, at 34 m from the road. Data from all measurement heights were used.

Both models underestimate concentrations in the regime of higher wind speeds and non-parallel wind directions; this underprediction is more pronounced for the GRAL model. The CAR-FMI model overestimates concentrations in low wind speed conditions, regardless of the wind direction, and in near-to-parallel wind conditions. This overprediction cannot therefore be entirely caused by the assumption of steady-state, homogeneous wind flow (for a perpendicular-to-road wind, this has only a minor influence). Both of these trends accumulate to produce relatively high inaccuracies in a regime with a lower wind speed and a nearly parallel wind direction. The performance of GRAL varies less in terms of the wind speed and direction; the model better simulates cases with low wind speed and wind direction nearly parallel to the road, as compared with the CAR-FMI model.

### 4.2.3 Nantes '99

The Nantes '99 experimental campaign took place in Nantes, France, and investigated the mechanisms of air pollution dispersion in a typical street canyon, at the same time providing a detailed unique data base to street-scale modellers for assessing their models.

The project was funded by the French Ministry of Environment within the inter-organism programme for air quality PRIMEQUAL-PREDIT. The experiment was driven as a cooperation between the Laboratory of Fluid Mechanics (CNRS - Ecole Centrale de Nantes), the Service Aerodynamics & Climatic Environment (CSTB), the CERMA Laboratory (CNRS - Ecole d'Architecture de Nantes), the regional air quality survey network Air Pays de la Loire, and the Service of Urban Environment of the city of Nantes. In addition two British scientists (Richard Griffiths, UMIST and Chris Jones, DERA Porton Down) measured very short distance tracer dispersion and transfer times.

### Major scientific issues investigated (Nantes '99)

The experimental campaign Nantes '99 was conceived as a first stage of the project URBCAP whose main aim was to determine the chemical and physical processes taking place within an urban canopy and affecting the local air pollution. Nantes'99 had four main objectives, namely the study of the wind field in the street, the determination of the production of turbulent kinetic energy induced by the motion of vehicles, the quantification of the influence of the distribution of street surface temperatures on the structure of the flow and consequently on the pollutant dispersion within the street, and the validation of several models developed by the teams participating in the campaign.

### Relevance to SATURN aims (Nantes '99)

One of the primary aims of the experiment was explicitly to provide data allowing to evaluate numerical simulations codes used for studying the dynamic and thermodynamic structure of the urban canopy layer, and to validate novel (sub-) models. More specifically three models were tested: SOLENE for the radiation-heat transfer and energy budget at the street surfaces (walls and ground), CHENSI and PHOENICS for the flow and turbulent diffusion field and for the transport-diffusion of pollutant tracers. The experiment aimed especially at investigating the level of turbulence generated by the traffic motion within the street. In addition, the intention of the experiment was to provide a reference data base for the further validation of street scale air quality models.

### Experimental set-up (Nantes '99)

The Nantes '99 experiment took place in June-July '99 in the centre of the city of Nantes (France). The "rue de Strasbourg" is a street canyon with approximately North-South orientation (332° from North). It is a high traffic one-way street, with three lanes, and a great homogeneity in buildings construction. The width of the street is 15 m and the mean height of the buildings is 22 m (H/W = 1.4). The street is 800 m long and the experimental section is located midway between two cross-roads which are 60 m apart.

The wind field and the vehicle induced turbulence were measured with sonic, 3-D propeller and hot wire anemometers at three levels on each side of the street. Air temperature and the temperature of the walls were measured using thermo-couples at the same levels as the anemometers. Carbon monoxide (CO) was chosen as the pollutant emissions tracer and it was also measured at three levels. Reference wind, CO concentration and temperature were measured on a small mast at 7 m over the roof. Radiation budget components (global, diffuse, and Infra-Red) were continuously measured over the roof. They were also measured at pedestrian level during one day, at several locations (on the sunny side, in the shadow). Traffic was measured by vehicle counters at eight different places within the street and within the lateral streets. Finally, the wind was measured at a height of 3 m, alternately in each of the transverse streets, with a light 3D bivane propeller anemometer.

During the experiment, a mobile laboratory was stationed 250 m South from the measurement section, measuring CO, $NO_x$, $SO_2$, dust and $O_3$ at 3.5 m from the ground, temperature and relative humidity at 5 m, and wind speed and direction at 10 m. CO and $NO_x$ were also measured by a permanent station of the air quality monitoring network 150 m North from the study section at 4 m from the ground.

At several occasions clusters of non-buoyant Helium-filled balloons have been released from pedestrian level: their times of residence were measured and their trajectories were recorded with a video camera.

The experiment emphasis being on the low wind conditions that are the most favourable for pollutant accumulation, the Intense Observation Periods (IOP) were defined as the days when wind speed was less than 3 m s$^{-1}$ at the reference level (30 m above the ground), with large solar radiation, and dense traffic. These conditions were chosen in order to correspond to high pollution episodes and the most effective thermal situations. During these IOP's, all the sensors were kept operational, otherwise, the sensors of the lowest levels were removed.

The Nantes '99 experiment has been complemented by a smaller experiment in September 2000, at the same site, mainly to document the momentum and heat transfer to the building walls with arrays of thermocouples and hot wire anemometers.

## Main results (Nantes '99)

A data base comprising the measurements taken during the experiment Nantes'99 has been constructed with Microsoft® Access. It includes only the IOP's. Two parts were defined within this data base: one data set called "Nantes'99" and another one called "Trafic'99" (Vachon et al. 1999, 2000a, b).

Each table of the "Nantes'99" data set corresponds to one full day of measurements (0:00 to 24:00 local time) with a time step of 15 minutes. It contains the measurements of wind direction, wind speed, fluctuations of the three wind components, temperature, and CO concentration at the different levels within the street, the measurements of pollution at the surrounding stations and the radiation components. The data base also includes wind measurements within the lateral streets with a bi-vane 3D propeller anemometer. In addition, the meteorological data obtained with the mobile laboratory are also included.

Each table of "Trafic'99" corresponds to a full day of measurements with a time step of 15 minutes for each traffic sensor. The flux of vehicles per lane, the total flux of vehicles and, according to the type of the counter, the mean velocity of the vehicles or the number of light and heavy-duty vehicles are included in the tables.

The data base is openly available on request for model validation, on one CD.

A detailed numerical model of the quarter buildings has been generated for the morphological analysis, for the numerical simulations and to build a physical model at the 1/20 scale. This model was thoroughly studied in the atmospheric wind tunnel of the University of Karlsruhe, for simulating the flows above and within the street and its neighbourhoods and it was equipped with mobile devices to simulate the generation of turbulence by the traffic.

The nine internal reports prepared contain among other the results on the local street-scale meteorology comparison to the regional meteorological survey, the evaluation of pollutant transfer times within the street from tracer balloon trajectories, a survey of alternate wind speed measurements obtained in the lateral streets around the experimental section, and the analysis of the flow simulations in the Karlsruhe wind tunnel.

Most of the major results obtained to date concern the validation of numerical models or sub-models. Interesting conclusions could be drawn from the comparison of the radiation and temperature measurements to the simulations with the model SOLENE. In particular, the surface temperature time evolution was found to be correctly modelled in conditions of radiation trapping, while it is being underestimated at sunshine periods due to a high sensitivity of the model to the surface albedo and to the convective heat transfer coefficient (Vachon 2001). Regarding the flow within the street in the presence of an important differential wall heating, it appeared that the thermal turbulent boundary layer at the wall is much thinner than expected, leading to an overestimation of the heat transfer at the wall by the numerical model (Louka et al. 2002). On the other hand, the investigation of the generation of turbulence by the traffic motion showed its influence on the pollutant dispersion and demonstrated its importance during low wind conditions, while novel modules were drafted and included in the model CHENSI and in the operational street quality model OSPM (Vachon et al. 2002; Berkowicz et al. 2002). Finally, it could be quantified to what extent thermal conditions and traffic induced turbulence affect the distribution of pollutant concentrations within the street as a function of time and meteorology, while it was demonstrated that it is important to properly position sensors for monitoring hot spot pollutant concentrations (Vachon 2001; Vachon et al. 2000a; Berkowicz et al. 2001).

### 4.2.4 Klagenfurt

Field measurements were undertaken in Klagenfurt in order to investigate pollution dispersion near tunnel portals. These measurements covered gaseous pollutants and meteorology over 4 months and a couple of tracer gas tests over a period of two days. The field experiments were supposed to provide data for model development and application.

### *Major scientific issues investigated (Klagenfurt)*

Road tunnels and covered roadways are becoming more and more part of a functioning urban road network. Policy is now to separate between traffic calming zones and a highly effective main street network. This leads to a shift of the pollution due to exhaust gases towards the two portal regions. For a big part of the whole region the air quality will be improved, but for those two areas, it will definitely become worse. In order to estimate the pollution concentration at the portal locations during the planning phase it is necessary to use appropriate dispersion models. Simple operational dispersion models are not designed to treat complex situations like junctions and tunnel portals. On the other hand complex models, which are capable to consider all the necessary details (buoyancy effects, exit velocity, building structures), can be operated only for single situations. They are not able to calculate concentration statistics, e.g. annual means, percentiles. There exists still a gap between simple – but operational - street models, and complex CFD models, which cannot be used in an operational way.

Field experiments were undertaken in order to investigate the special dispersion conditions in the vicinity of tunnel portals, and to develop and validate a new model for operational use. The field campaign included seven $SF_6$ tracer experiments over a time span of half an hour at a highway portal in flat terrain. During the $SF_6$ tests anticyclonic conditions prevailed. This allowed local wind systems to be developed. For most of the tests a large fraction of the wind speeds, recorded with the sonic anemometer, were below 0.8 m s$^{-1}$.

### Relevance to SATURN aims (Klagenfurt)

The main aim of the experiment was to provide a dataset for the evaluation of dispersion models for tunnel portals. The dispersion patterns show the distinct features of the jet and its interaction with the ambient air flow. The dataset was used in order to test different model types, which included simple empirical models up to numerical models. Finally an operational dispersion model was developed on the basis of a Lagrangian particle concept. The whole study leads to a better understanding of urban air pollution in the vicinity of tunnel portals. Especially near these vast emission sources the source-receptor relationship can be studied clearly.

### Experimental set-up (Klagenfurt)

*Meteorological data:* Meteorological measurements were performed in order to monitor the dispersion situations. As the region of the test site is well known for the frequent calm wind situations the usage of sonic anemometers was imperative, as otherwise no reliable meteorological information would be obtained. In addition to the sonic anemometers a standard cup anemometer was mounted on a 10m mast. The location of the meteorological equipment can be found in Fig. 4.5.

*Tracer gas experiments* were undertaken for investigating the behaviour of the tunnel air exhaust jet stream. SF$_6$ was used as tracer gas in order to obtain the parameters for short term exposure. 27 bag samplers were posted for getting

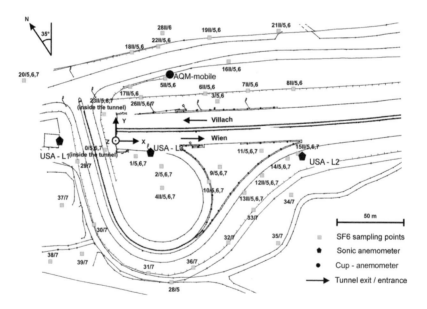

**Fig. 4.5.** Location of the meteorological equipment and the SF$_6$ sampling points

a clear picture of the exhaust jet of the tunnel portal. While one grab sampler per bore was installed inside the tunnel the remaining 25 were positioned depending on the meteorological conditions. The analysis of the $SF_6$ samples was made with an FTIR system using a White cell with a path length of 10 m. The detection limit was 3 ppb with an accuracy of +/- 5 %. Individual pumps with 0.3 l/min each were employed for the grab samplers. 10 l bags allowed a measurement time of 30 minutes. The 30 minutes period was chosen in order to have the same time intervals as the standard meteorological data. Fig. 4.5 shows the location of the sampling points for the experiments 1 to 4. The tracer gas was released some 1 km inside the tunnel for ensuring a well mixed exhaust air jet when exiting the tunnel. A mass flow controller was employed to control the $SF_6$ release.

### *Main results (Klagenfurt)*

The dispersion of pollutants from a roadway tunnel portal is largely determined by the interaction between the ambient wind and the jet stream leaving the tunnel portal. Due to the fact that the wind direction changed frequently it was possible to cover a broad variety of dispersion situations. The drawback of these varying winds was that the grab samplers were not always positioned at the most representative locations. In one case the wind direction changed by almost 180° shortly after the start of a test. This resulted in a number of zero samples. During the second test day the wind direction varied only by 90° which reduced the zero samples enormously. Only a few dispersion models appear in literature, which are designed to be used for regulatory purposes. These models are either empirical models, which may not be applicable for many different sites, or do not contain important physical effects like buoyancy phenomena. Here, a rather simple model has been developed, which takes into account most of the important physics playing a role in the dispersion process in the vicinity of tunnel portals. These are the jet stream, buoyancy, influence of ambient wind direction fluctuations on the position of the jet stream, and traffic induced turbulence.

Momentum of the exhaust jet (exit velocity): The exit velocity is a simple function of the piston effect of the traffic inside the tunnel, mechanical ventilation by fans, and the pressure gradient between the two portals. The velocity can be calculated according to the traffic piston equation.

Buoyancy: A temperature difference between the ambient and the tunnel air is caused by warming of the latter by the combustion process of the vehicles and a warming or cooling by heat transfer at the tunnel wall. Both will lead in most cases to a warmer tunnel air compared to the ambient one (except on summer days). Hence, a model not taking into account buoyancy will overestimate in most cases concentrations in the vicinity of roadway tunnel portals. Temperature differences typically can lie between 0 and 7 K.

Traffic induced turbulence: Especially heavy duty vehicles disturb the jet stream of a roadway tunnel and cause the jet stream to be orientated rather along the road compared to a "free" jet stream. The latter changes its direction more quickly towards the ambient wind direction.

Interaction of the ambient wind with the exhaust jet stream: The jet stream will change its direction more or less quickly according to the ambient wind direction. Because the ambient wind direction is changing over the considered time interval (in most cases 1/2 hour), the position of the jet stream will also vary. Thus, a model which computes the interaction between the jet stream and the mean ambient wind, without considering the wind direction fluctuations, will overestimate concentrations.

The meteorological conditions recorded and used for subsequent simulations can be seen in Table 4.1 where $U_a$ is the ambient wind speed, DD the observed direction, u* the friction velocity, L the Monin-Obukhov length, $\sigma_u$ the standard deviation of the along-wind and $\sigma_v$ of the cross-wind, $U_{pS}$ the tunnel air exit velocity, and $\theta_{S0} - \theta_A$ the temperature difference between tunnel and ambient air.

In most cases the mean wind speeds were quite low, which causes the ambient wind to meander (e.g. Etling 1990). In other words, the wind direction fluctuations are quite high in such conditions and the effect will not be negligible. The different wind directions during the experiments allow for a critical testing of the models' capability to simulate the position of the jet stream centreline correctly. In most cases the atmospheric stability was unstable, except for test case 4, where it was stable.

The new model shows the following overall performance. The fractional bias for all data (n=157) is very good and reads 0.012, the correlation coefficient of 0.75 is also satisfactory, and the percentage of concentrations, which were calculated within to a factor of two is 55 %. A linear regression analysis gives a slope of 1.11, and a shift of 8.8 µg m$^{-3}$. All statistical measures indicate the models ability to be used for engineering applications, were main emphasis is laid on the calculation of concentration statistics.

**Table 4.1.** Meteorological conditions during the field tests.

| # | $U_A$ (m s$^{-1}$) | DD (deg) | $u_*$ (m s$^{-1}$) | L (m) | $\sigma_u$ (m s$^{-1}$) | $\sigma_v$ (m s$^{-1}$) | $U_{pS}$ (m s$^{-1}$) | $\theta_{S0} - \theta_A$ (K) |
|---|---|---|---|---|---|---|---|---|
| 1 | 1.21 | 252 | 0.258 | -33.4 | 0.59 | 0.59 | 3.0 | 5.0 |
| 2 | 1.07 | 204 | 0.242 | -10.8 | 0.60 | 0.54 | 3.2 | 2.8 |
| 3 | 1.55 | 101 | 0.130 | -8.2 | 0.36 | 0.45 | 4.0 | 2.5 |
| 4 | 0.62 | 25 | 0.108 | 19.3 | 0.26 | 0.44 | 4.3 | 5.5 |
| 5 | 0.65 | 237 | 0.213 | -19.2 | 0.46 | 0.47 | 4.6 | 5.0 |
| 6 | 0.56 | 248 | 0.187 | -10.1 | 0.57 | 0.79 | 6.2 | 1.0 |
| 7 | 1.10 | 314 | 0.051 | -11.2 | 0.29 | 0.46 | 6.2 | 5.0 |

## 4.2.5 Podbielski Str.

The prevention of air pollution in traffic-loaded areas and the execution of the EU Directives ask for suitable tools for predicting pollutant concentrations. Different procedures and methods are operationally used to achieve that aim. Quality assurance asks to aspire to comparable procedures for concentration prediction. Ring tests comparable to tests with monitoring devices can essentially contribute to these comparisons. The Podbielski Strasse Exercise was used as a first attempt into this direction. The entire chain of the prediction process was checked, from the input parameters over the modelling of the dispersion to the predicted concentrations (Lohmeyer et al. 2002). The goal was to show the possible spread of the resulting concentrations predicted by different organisations/persons and/or different procedures if all users had the same input parameters at their disposal.

The meteorological parameters recorded were the wind direction and speed at the station HRSW, on the flat roof of the service building of the Lower Saxony State Agency for Ecology (NLÖ) in Göttinger Strasse in Hanover (see below), approximately 4 km off from Podbielski Strasse, 42 m above ground, 10 m of height over roof. Measurements of pollutant concentrations at the station HRSW were provided as hint for the background concentration (benzene, soot, $NO_2$, NO and CO as well as the 98-percentiles for $NO_2$, NO and CO). Traffic data were available (number of vehicles) between 06.30 and 18.30, providing mean traffic time series over the day, on weekdays, on Saturdays and on Sundays.

For benzene, a mean value was taken for ten months (April 1999 to January 2000) with 5 $\mu g/m^3$. The measured mean annual soot concentration was 3.1 $\mu g/m^3$. The measured annual mean for $NO_2$ was 50 $\mu g/m^3$ and for $NO_x$ 95 $\mu g/m^3$. The 98-percentiles amounted to 88 $\mu g/m^3$ for $NO_2$ and 247 $\mu g/m^3$ for $NO_x$. Based on these results and the meteorological input several models were applied and validated.

## 4.2.6 Göttinger Str.

The EU Air Quality Directives include air pollutant dispersion models as instruments of environmental politics. The quality of these models has to be checked. One part of that procedure is the validation, i.e. the comparison of the results of the models with specially designed and acquired reference data sets from field and wind tunnel measurements.

A data set for validation of microscale numerical dispersion models are compiled in the framework "Development and validation of tools for the implementation of European air quality policy in Germany" (VALIUM within AFO2000 of BMBF) co-ordinated by M. Schatzmann, University Hamburg.

Within an urban quarter around the "Göttinger Strasse" in Hanover, Germany, air pollutants and meteorological parameters are measured continuously at different sites by stations in the frame of the State Environmental Agency of Lower

Saxonian Monitoring System LÜN operating a permanent monitoring station. Apart from those long term measurements three intensive measurement campaigns with additional tracer experiments are planned. During the tracer experiments, a line source of approx. 96 m long operates along the median strip of the Göttinger Strasse. The width of the canyon is 25 m and buildings on both sides of the street are ca. 20 m high. It is a four-lane street canyon with a traffic load of ca. 30000 vehicles/day.

In August 2001, a pre-test was executed to check the experimental conception and set up. Samples of air were collected in bags at 12 locations within the street canyon and at the roofs of the buildings, and analysed later. The results show a reasonable distribution of the concentration in the street area. Later, the data will be compared to the results of wind tunnel experiments.

The first test of the experimental layout was, with minor reservations, successful. For the planned three intensive measuring campaigns, only insignificant modifications of the experimental set up will be necessary.


### 4.2.7 London

The London campaign investigated the variability of road-users (cyclists, bus and cars users) exposure to fine particulate matter ($PM_{2.5}$) for different routes, with its results having main application to Air Quality Management Systems (AQMSs).

The project was funded by the UK Engineering & Physical Science Research Council, under their Inland Surface Transport programme, with additional support from Go-Ahead Group p.l.c. and other transport operators, and collaboration with CERC Ltd.


### *Major scientific issues investigated (London)*

The campaign was designed to answer the following questions, with reference specifically to $PM_{2.5}$:

– To what extent does monitored or modelled fixed-point urban background or roadside air quality represent the level of air pollution people are exposed to while travelling on urban roads?
– How variable is road-user exposure to air pollution in Central London at the individual level, at the between-vehicle level, between different modes of transport (bus, car, bicycle), diurnally, from day to day, seasonally, and between different routes?
– What are the determinants of exposure?
– Is it possible for air quality management decision support software to include a facility to examine road-user exposure to air pollution, using existing Gaussian-type and street canyon dispersion modelling methodology?

– What would the implications for urban air quality management be if more attention was paid to road-user exposure instead of the current exclusive focus on fixed-point background or sometimes roadside concentration?

### Relevance to SATURN aims (London)

The aims of SATURN extend as far as the application of AQMS decision support software tools to assess population exposure to air pollution, usually by simple multiplication of modelled concentration by number of people living or working in each part of a city. The London exposure measurements go beyond SATURN's stated aims by considering exposure of moving receptors in the on-street microenvironment. This is an important contribution to the development of AQMS capability in this area, by allowing the appropriateness and accuracy of the simpler modelling techniques to be assessed. Insofar dispersion modelling might be capable of considering road-user exposure, the London results should also be incorporated into future AQMS development, if it is deemed to be useful to be able to consider such detail. Such AQMS development is one of the key integrating aims of SATURN (see Chapter 9).

### Experimental set-up (London)

Road user exposure to $PM_{2.5}$ fine particulate air pollution was measured using high-volume personal sampling equipment (Adams et al. 2001a) along 3 fixed routes within Central London during a 3-week summer 1999 and 3-week winter 2000 campaign, for the three modes of transport bicycle, bus and car. An additional summer 1999 cyclist exposure measurement campaign was made on variable routes corresponding to the volunteer cyclists' normal commuting journeys (Adams et al. 2001b). Each sample was analysed in the laboratory for total mass of particulate matter collected (including correction for humidity and pressure effects) and for blackness as a calibrated measure of the contribution of carbon principally from diesel exhaust (Adams et al. 2002). Simultaneous daily samples during the summer 1999 measurement period were taken from a fixed urban background monitoring site and analysed for major anions and cations by ion chromatography to quantify the contribution of long-range and regional transport of principally nitrate, sulphate and ammonium aerosol.

### Main results (London)

The data set collected is significantly larger than most, if not all, previous road-user exposure measurement data sets. After quality control, the number of valid samples from the fixed-route summer campaign was cyclist: 40, bus: 36, car: 42; for the fixed-route winter campaign cyclist: 56, bus: 32, car: 12, plus 120 variable-route summer cyclist samples.

Time-series of modelled or measured roadside or urban background air quality correlate well with road-user exposure. The road-user exposure, however, is more variable than roadside concentration and twofold the urban background concentration.

A large fraction of the exposure variability is at the individual and between-vehicle level, with day-to-day and between route variability also being very important. For example, the difference in exposure between two cyclists travelling on the same route at the same time is similar to the difference in exposure of one of those cyclists on two different routes or two different days. Above the noise of the other sources of variability however, it was not possible to detect any systematic difference between the average exposure of several cyclists and the average exposure of several car or bus users on the same route and the same day. Most of the exposure variability at individual level and between routes is attributable to the black carbon fraction of the $PM_{2.5}$, while the non-carbon fraction was more important in the day-to-day variability. For the black carbon part of $PM_{2.5}$, it was also possible to detect almost a factor of two differences between modes of transport, with car drivers having greater exposure than cyclists.

Regression modelling (Adams et al. 2001c) shows that the main determinants of exposure are wind speed and route. Detailed information about instantaneous position on road was not available to be included in the regression modelling, so the possibility that this is a cause for the unexplained individual-level variability is left as a plausible hypothesis to be tested by further research.

The ADMS-Urban dispersion modelling system is capable of reproducing the dependence of exposure on wind speed and route, including the day-to-day variability where sufficient information about imported secondary transboundary $PM_{2.5}$ is available for inclusion in the model input. It is not, however, currently capable of accounting for the individual-level variability in exposure, nor all the enhancement of road-user exposure relative to roadside air quality. The average measured exposure is similar to the modelled concentration at the most polluted part of the route, not the average along the route assuming steady movement at constant speed.

One insight gained by our approach to modelling road-user exposure is the large difference between the most polluted 10% of a route and the less polluted 90%. Action to tackle the hot-spots is likely to result in the most effective improvement of the quality of the urban environment, including such simple measures as decreasing delays to cyclists and bus users at congested locations and physical separation of non-polluting road-users from polluting vehicles. The source apportionment of $PM_{2.5}$ at the exposure hot-spots is also quite different to that along the rest of the route. At the hot-spot, emissions from the nearest road dominates. Along the rest of the route, the contribution of imported secondary transboundary particulate matter is the same order of magnitude as that of the local emissions. Local authorities, who are often the main end users of the AQMSs that SATURN is developing, therefore have more control over the hot-spots and exposure than they do over general urban air quality. This research continues and will be integrated with street canyon intersection modelling and wind tunnel studies. Major on-street measurement campaigns are planned for Spring 2003 and 2004 at the junction of Marylebone Road with Gloucester Place.

# 4.3 Urban scale campaigns

### 4.3.1 LisbEx – Experimental studies in the Portuguese Atlantic coast

The Great Lisbon Area is, due to its industrial and urban importance, an example of a region with high emission levels and photochemical air pollution. On the other hand, it is located on a coastal zone with a complex coastline associated to significant terrain features and sea/land breezes circulation, which result in a complex airflow circulation. Two summer campaigns, in 1996 and 1997, were dedicated to the investigation of the effects of flow circulation on pollutants transport and the model evaluation.

### *Major scientific issues investigated (LisbEx)*

The knowledge and characterisation of mesoscale airflow patterns as well as its effect on the dispersion of atmospheric pollutants in coastal areas is fundamental. The objectives of the Lisbon campaigns were the characterisation of meteorology and air quality situation during summer and focused on the study of the breeze circulation in coastal zone, vertical structure of atmospheric boundary layer, formation and transport of photochemical pollution, and model evaluation. The application of numerical modelling as a complementary tool to experimental fieldwork was essential for this study.

### *Relevance to SATURN aims (LisbEx)*

The LisbEx databases are fundamental for model development particularly applied to coastal zones, and its validation contributing to a better performance of models. The Lisbon campaigns also represented an effort on monitoring the Great Lisbon Area air quality showing the need of improvement of the air quality network in terms of spatial coverage.

### *Experimental set-up (LisbEx)*

Two field studies (meteorology and air quality) were carried out in Lisbon coastal area from 8 to 18 July 1996 and 1997 (LisbEx 96 and 97), searching for typical synoptic summer situations.

The Lisbon campaigns were structured in order to integrate all monitoring stations (public or private) located on the study domain (area about 200 km × 200 km). Mobile monitoring stations with locations based on numerical simulation were also used. Generally, temperature, relative humidity, wind speed and direction were measured at every meteorological station. $O_3$, CO, $NO_x$ and $SO_2$ concentrations were measured at air quality stations (Fig. 4.6). Besides, ground-based information, radiosondes and tether-balloon soundings have been done. The tether-balloon soundings system includes both meteorological data acquisition and $O_3$ sensor. In addition, satellite images, forecast maps, etc were available from the Portuguese Meteorological Institute (Borrego et al. 1998).

In the summer '97 campaign, the vertical structure of the atmospheric boundary layer was also studied with aircraft measurements in the frame of the STAAARTE Programme (Scientific Training and Access to Aircraft for Atmospheric Research Throughout Europe). The flights performed with the instrumented aircraft had "in situ" monitoring of atmospheric pollutants.

The synoptic situation during LisbEx 96, according to the mean sea-level pressure analysis based on the 12:00 UTC synoptic charts from the Portuguese Meteorological Institute, was dominated by the presence of the Azores anticyclone centred over the Northwest of Iberian Peninsula. Two prominent synoptic meteorological patterns were identified. The first pattern, from 8 to 15 July, was characterised by an eastern wind, with hot and dry air, transported in the Azores anticyclone flow that extended in ridge to Northeast. In the second pattern, during 16 and 17 July, the synoptic situation was influenced by the joint action of an anticyclone centred over Scotland and an under-pressure valley over Iberian Peninsula that generated a Western flow over Portugal. During LisbEx 97, the meteorology was conditioned by the presence of the Azores anticyclone that was either centred over the Azores islands or at the West or Northwest extending in ridge to France. Also, the presence of depressions on the Iberian Peninsula and British islands resulted in atmospheric instability, being observed the approximation and transit of a cold front, followed by thunderstorm and showers.

### Main results (LisbEx)

Based on their locations, the stations were divided in coastal, inland and urban stations. In LisbEx 96 period, the mean temperature at coastal stations (Cabo da Malha, Cabo Carvoeiro and Sines) was 20 °C, the maximum being observed at Cabo da Malha (32,8 °C) whilst for the urban stations (Lisbon, Barreiro, Setúbal and Tires) the maximum was registered at Tires (38 °C). The mean temperature of 25 °C was found for all urban locations except Setúbal (21 °C). The air temperature was generally higher for the inland stations with a maximum of 39 °C observed at Pego. At the different monitoring sites the same pattern for the wind speed variations was observed, but it was generally lower at inland sites. The average wind speed at the coastal stations was about 4 m s$^{-1}$ while at others it was about 3 m s$^{-1}$.

The air temperature during LisbEx 97 was generally lower than the LisbEx 96 campaign with the maximum observed at inland sites (Évora and Beja - 33 °C). The maximum wind speed at the coastal stations varied between 5 and 12 m s$^{-1}$ while at others stations varied it between 4 and 8 m s$^{-1}$ (Borrego et al. 1999).

The atmosphere vertical structure was studied up to 8000 m with vertical profiles for pressure, temperature and wind speed and direction obtained from radiosondes realised at Lisbon station. Fig. 4.7 presents the vertical profiles for 10 July 1996. This day was characterised by a synoptic forcing from N-NE with moderate wind speeds at the surface. The predominant flow from N-NE over Portugal is representative of a typical summer situation.

During the Lisbon field campaigns O$_3$ was measured at nine stations (Fig. 4.6). In Table 4.2 the mean, maximum, minimum and 98$^{th}$ percentile of hourly average O$_3$ concentration obtained at the various stations are presented. It shows that the higher O$_3$ concentrations were observed at the stations located far from the urban centres (Pego, Monte Velho and Monte Chãos). This situation was expected as O$_3$ is depleted by high concentrations of NO$_x$ caused by emission of heavy traffic in the urban centres (Valinhas 1998).

The diurnal O$_3$ concentrations at the nine measuring stations in GLA generally showed the typical pattern of photochemical smog formation, namely, a steep rise in O$_3$ concentrations in the first morning hours at 8:00 to 9:00 UTC with a maximum in the first afternoon hours around 14:00 UTC.

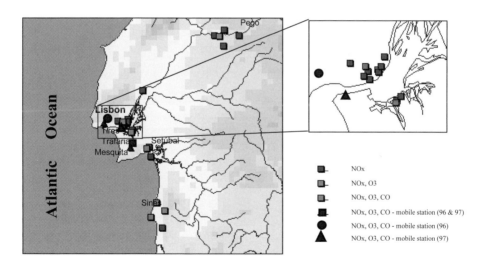

**Fig. 4.6.** Location of air quality stations and pollutants measured at Great Lisbon Area during LisbEx

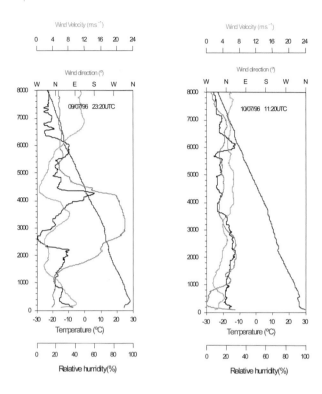

**Fig. 4.7.** Radiosondes realised at Lisbon station during LisbEx on 10 July 1996

**Table 4.2.** Mean, maximum, minimum and 98th percentile of hourly average $O_3$ concentration ($\mu$g m$^{-3}$) during the LisbEx field campaigns

|  | Mean | | Max. | | Min. | | 98th percentile | |
|---|---|---|---|---|---|---|---|---|
|  | 96 | 97 | 96 | 97 | 96 | 97 | 96 | 97 |
| Pego | 84 | 78 | 173 | 107 | 19 | 7 | 199 | 98 |
| Rua do Século | 38 | 32 | 61 | 77 | 27 | 2 | 52 | 65 |
| Entrecampos | 8 | 18 | 23 | 50 | 2 | 0 | 19 | 43 |
| Tires | 72 | | 168 | | 2 | | 155 | |
| Trafaria | | 63 | | 108 | | 15 | | 104 |
| Hospital Velho | 39 | 67 | 192 | 143 | 20 | 14 | 182 | 112 |
| Monte Chãos | 49 | 72 | 192 | 218 | 18 | 16 | 160 | 193 |
| Monte Velho | 107 | 69 | 155 | 116 | 65 | 16 | 145 | 110 |

## 4.3.2 Graz

The air quality investigation in the city of Graz included two major measurement campaigns in winter 98/99 and summer 99. Two episodes were investigated during high pressure periods over Central Europe under which local wind systems could develop. It was shown that air quality is affected by the location of the city in the southeast of the Alps in a basin surrounded by mountains. The existence of a counter-current over the city area was investigated together with its dependence on stable atmospheric conditions frequently occurring during night and morning hours. The implications to pollution dispersion were illustrated.

### *Major scientific issues investigated (Graz)*

The research project DATE Graz (Dispersion of Atmospheric Trace Elements taking the city of Graz as an example) aimed at the investigation of mesoscale pollution dispersion of a city in complex terrain in a pre-alpine region south of the Alps. Two episodes have been investigated which were characterised by anticyclonic fair weather conditions. Main objectives of the measurement campaigns were dedicated to the influence of local wind systems and strong temperature inversions on air quality; the vertical structure of the boundary layer and its temporal development; the comparison of point measurements against open path air quality measurements; and the validation of models developed by the teams participating in the project.

### *Relevance to SATURN aims (Graz)*

The measurement campaign for the City of Graz was designed to investigate the dispersion conditions of cities situated in valleys and basins in pre-alpine regions during anticyclonic weather conditions. Graz is well suited for such an investigation as it is situated in the south-east of the Alps in the transition area of mountainous to flat land. The city itself is located in the valley of the river Mur, which forms a basin surrounded by small mountains. A major influence on the air quality of Graz is the development of local wind systems which interact with larger circulation patterns. The local wind systems consist of the katabatic flows from the small tributary valleys in the east of the city and the frequent development of stable low wind situations in the basin. These local systems are overlapped and interact with the mountain-valley wind system of the valley of the river Mur. This valley enters the basin from north-west and origins from some 190 km northwest in the Alps. It produces a distinct valley wind at two to three hours after sunset. Nevertheless it is not able to enter the basin down to its ground due to strong temperature inversions in the basin. The measurements show a counter-flow from south going underneath the valley wind from northwest. Numerical investigations show that this is very probably a Froude-number dependent flow circulation around the small mountain ridge in the west of the city. The relevance to SATURN was to show the influence of the mountainous topography on the air quality of cities. Although Graz is a rather small European city, the air quality during the described conditions is poor. The measurement campaign aimed at the investigation of the different influences on air quality and resulted in two measurement data sets.

Source-receptor relationships in Graz during the winter episode were studied with a mesoscale dispersion model. Input for the simulation was the emission inventory and extensive meteorological measurement data. Results of the simulation reflect the distinct patterns of daily variations of air quality levels measured. The influence of meteorology, emission patterns and chemical reactions is evident and can be qualitatively and partly quantitatively simulated by the model. A validation attempt was made using air quality data from the monitoring network. Several papers describe the measured phenomena also in comparison to the model results.

## *Experimental set-up (Graz)*

The area investigated is characterized by a continuous transition from the Alps to flatlands. The major part of the city is located in a basin surrounded by small mountains to the west and north (Fig. 4.8). From the east, tributary valleys enter the city basin coming from a small north-south oriented ridge, which forms a barrier approximately 10 km to the east. To the south the oval basin (25 km S-N and 15 km W-E) is surrounded by small hills. The river Mur enters the basin in the north-west through a narrow gap, and leaves it in the south. The contour lines indicate the altitude in 50 m intervals. The highest elevation is at a height of 1450 m a.s.l.. The position of the area within Europe is indicated in the small square in the lower part of Fig. 4.8. The Alps are located northwest of this area. The measured quantities are shown in Table 4.3.

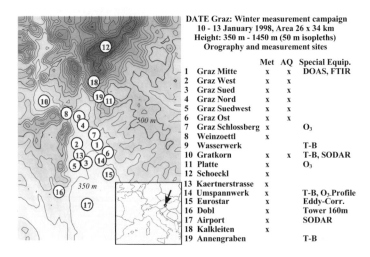

**DATE Graz: Winter measurement campaign**
**10 - 13 January 1998, Area 26 x 34 km**
**Height: 350 m - 1450 m (50 m isopleths)**
**Orography and measurement sites**

|    |                 | Met | AQ | Special Equip. |
|----|-----------------|-----|----|----------------|
| 1  | Graz Mitte      | x   | x  | DOAS, FTIR     |
| 2  | Graz West       | x   | x  |                |
| 3  | Graz Sued       | x   | x  |                |
| 4  | Graz Nord       | x   | x  |                |
| 5  | Graz Suedwest   | x   | x  |                |
| 6  | Graz Ost        | x   | x  |                |
| 7  | Graz Schlossberg | x  |    | $O_3$          |
| 8  | Weinzoettl      | x   |    |                |
| 9  | Wasserwerk      |     |    | T-B            |
| 10 | Gratkorn        | x   | x  | T-B, SODAR     |
| 11 | Platte          | x   |    | $O_3$          |
| 12 | Schoeckl        | x   |    |                |
| 13 | Kaertnerstrasse | x   |    |                |
| 14 | Umspannwerk     | x   |    | T-B, $O_3$ Profile |
| 15 | Eurostar        | x   |    | Eddy-Corr.     |
| 16 | Dobl            | x   |    | Tower 160m     |
| 17 | Airport         | x   |    | SODAR          |
| 18 | Kalkleiten      | x   |    |                |
| 19 | Annengraben     |     |    | T-B            |

**Fig. 4.8.** The location of the Graz campaign

**Table 4.3.** The measured quantities in Graz experimental campaign

| Met.: | Meteorological standard devices (wind speed, wind direction, temperature) |
|---|---|
| AQ: | Air quality monitoring sites (NO, $NO_2$, $SO_2$, TSP, CO) |
| DOAS: | Differential optical absorption spectroscopy |
| FTIR: | Fourier transformed infrared spectroscopy |
| T-B: | Tethered balloon measurements (hourly) |
| $O_3$-Profile: | Hourly vertical $O_3$-profiles |
| Eddy-Corr.: | Eddy correlation measurement by sonic anemometer |
| Tower: | Meteorological tower with met. standard devices at 10 m, 30 m 160 m |

The winter measurement campaign was designed to study a typical winter smog situation. The first anticyclonic weather episode during that winter started on January 9. The anticyclone in the 500 hPa ridge was very intense and prevented fog by the downdraft of dry air even at ground level in the area under investigation. The episode ended with the advection of ground fog in the late evening of January 12. The measurements started at 6 p.m. of January 9 and ended after 82 hours in the morning of January 13. During the three days of the field experiment, hourly vertical profiles of wind speed and direction, temperature and humidity were measured at four locations (indicated by T-B in Fig. 4.8) in the chosen area. In addition, one meteorological tower, two SODARs, one eddy correlation measurement device and at least 15 meteorological standard devices on the ground were in operation. At the same time 1 DOAS system and 7 air quality stations for the measurement of standard pollutants were in operation.

## Main results (Graz)

Three typical local wind systems can be distinguished for the city of Graz (Almbauer et al. 1995, Lazar and Podesser 1999): (a) Mountain-valley wind system; (b) Drainage flows from the tributary valleys east of Graz; (c) Cold air from the southerly basin.

Measurements of the winter campaign showed the expected local wind systems. The mountain valley wind system of the river Mur is clearly documented in the observed vertical wind profiles south (Fig. 4.9) and north (Fig. 4.10) of the city centre. The arrows in the diagrams show the hourly measured horizontal wind direction and wind speed up to a height of 500 m above ground level. Approximately two hours after sunset at 18:00 LST (Local Standard Time) the mountain wind entered the basin of Graz. It reached ground level on January 10 and 11 in the north of the city centre at site 9. South of the city centre southerly wind directions remained even at the beginning of the valley wind. During the night the southerly flow developed and increased in depth up to a height of approximately 100 m - as can be seen in Figs 4.9 and 4.10. Above, the low level jet of the mountain wind with velocities up to 8 m s$^{-1}$ remained during the night and ended in the late morning. The centre of the low level jet moved to higher altitudes during both

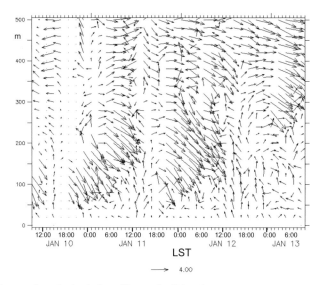

**Fig. 4.9.** Measured vertical wind profile south of the city

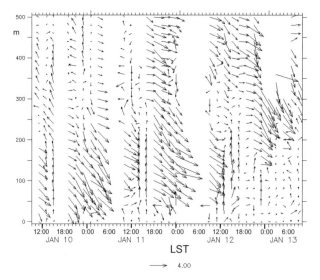

**Fig. 4.10.** Measured vertical wind profile north of the city

nights. During the day a weak valley wind from the south developed. The depth of the valley wind correlated well with the depth of the isothermal layer.

The measured time height diagram of $O_3$ (Fig. 4.11) and temperature (Fig. 4.12) at site 14 indicate the build up of a well-mixed layer during all three days (January 10 – 12) between 11:00 and 15:00 LST. During the night an inversion layer with a growing height was measured. Differences between the situations North and South of the city are explained in detail in Piringer and Baumann (1999). Differences are accounted for by the influence of the increased roughness length in the city on the mountain wind.

Air quality results are presented for NO and $NO_2$ as they are the pollutants which are closest to exceeding present threshold values. NO concentrations are influenced by: (a) the daily and hourly variation of emission rates depending on traffic and industry activity; (b) the complex flow field determined by the local wind system; (c) chemical reactions with importance for the conversion of NO to $NO_2$.

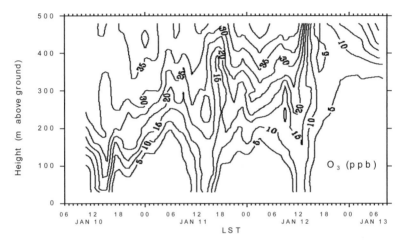

**Fig. 4.11.** Measured $O_3$ at site 14

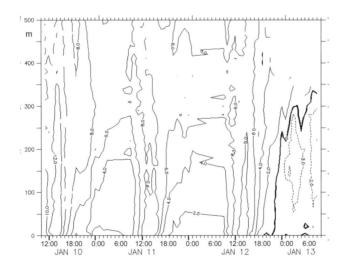

**Fig. 4.12.** Measured temperature at site 14

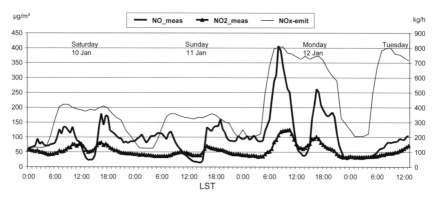

**Fig. 4.13.** Measured concentrations of NO, NO₂ and NOₓ emissions

In order to explain in general the daily variations of $NO_2$-concentrations, the mean NO and $NO_2$-concentrations for the whole city were calculated from sites 1, 2, 3, 5 and the DOAS records (Fig. 4.13). NO concentrations during both nights from January 10 to 11 and from January 11 to 12 are on a level of approximately 100 µg/m³. $NO_2$ concentrations decrease from approximately 75 µg/m³ in the evening to 40 µg/m³ in the early morning. On Monday a strong increase over time, up to daily maximum of more than 400 µg/m³ at about 8:00 LST was recorded. $NO_2$ concentrations rise slowly after sunrise with a Monday-maximum of more than 120 µg/m³, three hours after the NO maximum at about 11:00 LST. The fast conversion of NO to $NO_2$ is caused by the down-mixing of $O_3$ from higher altitudes (Fig. 4.11). The $O_3$ concentration during all three days in the residual layer above the inversion is about 30–40 ppb. All three days show a distinct minimum NO concentration at noon with values less than 40 µg/m³. $NO_2$ decreases more slowly to values between 40 to 60 µg/m³. During all three days NO-concentrations increase shortly after the breakdown of the well-mixed layer at 15:00 LST. On Saturday and Sunday the maxima in the afternoon are higher as a result of increased traffic. On Monday the second NO maximum is lower. $NO_2$-concentrations show the same behaviour as NO for their maxima on all three days. The sum of NO and $NO_2$ concentration decreases after 19:00 LST in line with the normal reduction in emissions. NO-concentration shows a good co-incidence with predicted emission rates. Vertical soundings of $O_3$ for the three days confirmed the assumption of $O_3$ down-mixing (Fig. 4.11). During the first two days, the $O_3$ values in the lowest 150 to 200 m increase with a rate of about 5 ppb per hour between 11:00 and 15:00 LST. The maximum values reach up to 20 ppb. During the rest of the days the $O_3$ concentrations were at a level less than 1.5 ppb near ground. Daily mean values of $O_3$ concentrations at the monitoring stations 2, 4 (at the city ground level) and 11, at a height of 210 m a.g.l. show a strong vertical stratification. The temperature inversion and weak winds prevented an exchange of air masses.

The mesoscale dispersion model GRAMM was applied for the simulation of air quality during the winter episode. Instead of using a nesting procedure initial and boundary conditions were directly obtained from the numerous observations within the model domain by objective analysis.

A prerequisite for predicting air quality is to simulate the complex flow fields correctly. The dominant mountain-valley wind system of the Mur valley with its characteristic diurnal change in wind direction from north-westerly winds in the night to daytime southerly winds is reproduced well in the simulation. Simulated wind speeds of the up-valley flow agree well with observations. On the contrary, the local drainage flows on the eastern side of Graz were not reproduced by the mesoscale model. The modelled temperature agreed well with observations during night-time hours, where an inversion layer with a temperature increase of 6 K within the first 250 m above ground level developed. During the day the inversion broke up and an isothermal stratification was observed. The warming of the air after sunrise and also the cooling in the afternoon is somewhat slower in the simulation.

Concerning the simulated NO and $NO_2$ concentrations, the strong increase during the morning rush hours on Saturday and Monday is well reproduced by the model. In particular, results in terms of the $NO_2$ levels agree fairly well with the measurements. The agreement is less satisfactory regarding individual peaks for Saturday morning and evening and the Monday morning rush hour. Results for the other sites show a similar behaviour. At most sites NO concentrations are underestimated. The reason for this might be either too low emission rates or wrong dispersion conditions.

## 4.3.3 Milan

The urban areas in Northern Italy are characterised by significant photochemical pollution associated with the geographical, meteorological and emission systems complexity. Urban experimental campaigns in the Lombardia region were designed and managed in order to improve knowledge of the urban dispersive and photochemical properties. Pollutant levels were investigated, with emphasis on photochemical pollution and particulate matter.

### *Major scientific issues investigated (Milan)*

The experimental campaigns in Milan aimed at characterising the chemical regimes near and inside the city plume, improving the knowledge of the urban dispersive and photochemical properties during the cold season, improving the vertical chemical characterisation, characterising the organic and inorganic mass composition of $PM_{10}$, and determining the urban aerosol mass closure and the relation between secondary aerosol formation and photochemical properties.

## Relevance to SATURN aims (Milan)

The experimental campaigns in Lombardia region were devoted to measure meteorological parameters and chemical compounds in order to provide data for model validation, as well as to develop emission inventories necessary for model application in the frame of different EUROTRAC-2 subprojects, while a "Milan focal point" task force was established in summer '98.

## Experimental set-up (Milan)

Seven experimental campaigns were performed in the years 1997 – 2001 devoted to measure meteorological parameters and chemical compounds at ground level and for vertical profiles. The Research Institutes involved in the campaigns were CESI, ENI Tecnologie and Università di Torino.

Two summer experimental campaigns were organised in 1997. During June-July, the measurements in rural and suburban areas of Milan were performed in order to characterise chemical regimes near and inside the city plume. From July until September the data were collected in Turin using sonic anemometer and SODAR.

During spring 1998 measurements were performed in several places: Milan (Piazza Carbonari, Viale Marche, Segrate and Turbigo Bresso suburban site); Bresso airport and Turbigo (ENEL thermal power plant); and Redecesio (suburban site – East Milan). The Doppler SODAR and RASS sonic anemometers, hydrometers and fast ozonometers were used during the campaign.

During the winter 1999 experimental campaign in the Milan area, ground level and vertical profile measurements of meteorological and chemical variables were measured. Part of the winter campaign took place in particular, from 01/02/99 to 25/03/99, at Carbonari Square, in the northern part of Milan urban area, near the main railway station. It was mainly devoted to improve knowledge in urban dispersion and photochemistry properties during the cold season, and to extend and complete the data set started with the summer campaign (summer 1998). According to previous experiences, a relevant experimental effort was performed in order to improve the vertical chemical characterisation. Beyond a standard ground level meteorological and chemical characterisation, the winter campaign had, as a special focal point, the measurement of vertical pollution structure related to meteorological conditions.

The experimental activity in July 2000 was mainly focused on urban aerosol chemical composition and aerodynamic distribution investigation. Aerosol analysis was combined with a complete characterisation of the primary and photochemical pollutants both at ground level and in the vertical structure. In January 2001 and in the period September-October of 2001 a mobile laboratory was used for measurements in Milan (Piazza Carbonari).

## *Main results (Milan)*

An extensive experimental work was performed in Milan. The databases included:

- Meteorological parameters, i.e. conventional ground level meteorological measurements; vertical profiles of wind, turbulence and temperature in the Planetary Boundary Layer; sensible and turbulent heat fluxes and momentum at different levels.
- Chemical compounds, namely, NMHC C2-C9, HC, BTEX, $CH_4$, aldehydes, carbonyl compounds, $PM_{10}$, $PM_{2.5}$, CO, $NO_x$, $O_3$, $H_2S$, $SO_2$, $HNO_3$, $HNO_2$ at ground level; vertical profile of $O_3$ and aldehydes; aerosol composition (ammonium, nitrate and sulfate, OC, EC, anions and cations) divided in different aerodynamic classes.

The experiments in Milan intercompare the instrumentation among different groups and improved the measurements quality control and assessment. The collected data were used for comparison with CALMET model prediction contributing for critical review of methods for estimating mixing height, turbulence parameterisation and vertical diffusivities in urban areas. The results helped to understand deeper which chemical multi-phase modelling schemes should be implemented in the mesoscale models.

### 4.3.4 Marseille, the ESCOMPTE and UBL/CLU experiments

The ESCOMPTE programme (Experiment on Site to Constraint the Models of atmospheric Pollution and Transport of Emissions) aimed at improving the knowledge on the meteorological and chemical conditions prevailing during photochemical episodes. The ESCOMPTE program included the (a) experimental campaign, (b) the construction of the detailed emission inventory over the whole region, (c) the construction of a comprehensive and extensive data base to be placed at the disposal of the scientific community.

UBL/CLU-ESCOMPTE was an "associated project" of ESCOMPTE, with an additional experimental set-up to document the fine scale dynamics and thermodynamics of the urban atmosphere over the Marseille area, situated at Mediterranean coast, and involving more than one million of inhabitants.

## *Major scientific issues investigated (ESCOMPTE and UBL/CLU)*

In order to be able to take in advance decisions limiting the amplitude and the effects of photochemical episodes requires, for the pertinent space and time scales, efficient prediction tools to warn the public and to limit the health impact, and deterministic simulation models to assess the influence of the pollution reduction decisions by the public authorities. The simulations require four main modules: the evaluation at any time of the natural and anthropogenic emissions, the computation of the transport-diffusion in the atmosphere, the computation of the chemical transformations for the gases and particulates, the deposition of the oxidized species on surfaces and vegetation. The ESCOMPTE programme aimed at providing a reference data set allowing to validate these modules in detail.

The chosen method included on the one hand the construction of a detailed emission inventory of fixed and mobile sources, at the scale of 1km×1km×1h, and on the other hand to conduct a field experiment allowing to measure at the ground, at sea, and in altitude the chemical composition and the dynamic and thermodynamic characteristics of the atmosphere from the local scale to the regional scale. The aim of the experiment was to obtain all the data necessary to constraint and/or to qualify the above mentioned modules, at coherent space and time scales. The plan left room for side projects taking advantage of the general set up and also participating in the main objective.

The UBL/CLU-ESCOMPTE experiment aimed at documenting the four-dimensional structure of the Urban Boundary Layer in connection with the urban canopy thermodynamics during a 7 weeks summer period of low wind and breeze conditions. The objective was mainly to construct a data base allowing to test urban energy exchange schemes and high resolution meteorological and chemistry-transport models.

The project took advantage of the large experimental set-up of the campaign ESCOMPTE over the Berre-Marseille area, especially as concerns remote sensing from ground, airborne measurements, and the intense documentation of the regional meteorology.

### *Relevance to SATURN aims (ESCOMPTE and UBL/CLU)*

Both experiments are directly aiming at validating models that are required for understanding pollution episodes and predicting air quality. While ESCOMPTE focuses on the $O_3$-VOC photochemistry, at the regional scale during several diurnal cycles for various sunny meteorological conditions, UBL/CLU-ESCOMPTE focuses on the urban atmosphere with highly inhomogeneous distributions of urban land use and pollutant sources. Both experiments document densely the atmospheric dynamics at the pertinent spatial and temporal scales, allowing to test and validate transport-diffusion models. UBL/CLU documents more extensively than ever the urban canopy thermodynamics, allowing to test and validate the urban energy models. ESCOMPTE documents more extensively than ever the gas and particulate concentration distributions and the photo-chemical transformations, allowing to test the tropospheric chemistry models. At the urban scale the documentation of pollutant concentrations was ensured by the permanent network of the air quality survey service AIRMARAIX.

## The experimental set-up (ESCOMPTE and UBL/CLU)

The selected site for ESCOMPTE campaign is a domain of about 100 km by 100 km around the Marseille conurbation and the area of Fos-Berre. In this area the highest $O_3$ episodes of all the French territory have been registered. The pollutant sources are multiple, and include urban sources within the Marseille conurbation and industrial within the Fos-Berre complex. The geography is complex, with land-sea interactions through a twisted coast. This means that the modelling exercise following the experiment will be a demanding test for the models, but with quite well identified dynamic features (thermal contrasts and orography influence).

The ESCOMPTE set-up

The campaign ran from June 4 to July 17, 2001 and 5 Intense Observation Periods of 3 days average were selected for complete deployment of the sensors and flights of the instrumented aircrafts (see URL 4.1). The ground-based platforms, specifically deployed for the experiment, involve 20 stations, equipped for gas ($O_3$, $NO_x$, VOCs,) and/or particles measurements; among them, two were installed on ships, and two mobile stations were placed according to the plume locations. The surface energy budget was measured at 9 sites to cover the landscape variety in the area; among them, four were in the urbanized area of Marseille (see below); at some sites, fluxes of trace gases ($O_3$, $NO_x$) were also measured for emission/deposition velocities computation. The meteorological basic parameters (wind, temperature, moisture and radiation) were measured on the 9 abovementioned and on 5 complementary sites. Wind profile was continuously measured on 12 sites by 7 sodars, 4 UHF and 4 VHF radars. A scanning Doppler Lidar measured the 3-D wind field over the Marseille agglomeration. 3 upward pointing, and 2 scanning $O_3$ lidars were set-up on a SW-NE axis (main breeze axis) through the domain. 4 radiosonde systems, 2 of them capable of $O_3$ profiling, were activated during the IOPs. The pollutants plumes were then tracked by 33 constant-volume balloons, launched in the boundary layer from the emission areas (Marseille city or Berre pond); they were equipped with radiosondes, some of them with an $O_3$ probe. 7 aircraft were flown during pollution episodes ; 4 (DO 128 from IMK, FRG ; Fokker 27 from INSU, France ; Merlin 4 and Piper Aztec 23 from Météo-France) were able to in situ document dynamics and chemistry ; a ULM from IFU (FGR) measured $O_3$, aerosol and UV radiation ; the Falcon 20 from DLR (FRG) embarked the Doppler lidar WIND ; and a Piper Aztec 28 embarked an IR camera for surface temperature characterization over the Marseille area (Mestayer and Durand 2002).

The UBL/CLU set-up

The UBL/CLU instrumentation was mainly deployed at 5 sites along the North-South axis of the city, roughly parallel to the shoreline. Four urban sites were equipped with micro-meteorological masts raising some 12 to 20 m above the urban canopy level, where all the turbulent and radiation fluxes necessary to monitor the canopy surface energy budget were continuously measured. The turbulent fluxes were measured at 2 levels, except at the Observatoire site at only one level at 12 m above ground. The central site (CAA) was located in the rather uniform, 19th Century, dense part of the city center. It was also equipped with an array of up to 19 IR radio-thermometers, either fixed to monitor the surface temperature of selected elementary surfaces, or hand-held to evaluate surface temperature distributions during some periods of intense observation. In this urban fragment thermometers also monitored the heat exchanges between building inside and outside during some periods. Two IR radio-thermometers were also operated at the North site located in a suburban area of mixed constructions to monitor the composite surface temperatures of the ensembles immediately North and South of the site.

The two sub-urban sites (GLM and St Jerome) were equipped with mini-SODARs sounding the atmospheric surface layer while the fourth urban site (Observatoire), close to the city center was equipped with a wind profiler UHF radar and a tethered balloon occasionally measuring thermodynamic and $O_3$ profiles from 20 to 300 m. Two scintillometers were set to measure the integrated heat flux over the city center, with 2.5 km optical paths oriented N-S and E-W. At the hilly northern border of the city, the site Vallon Dol hosted a 10 m high mast with a sonic anemometer, a RASS-SODAR vertical sounder, and two 3-D scanning LIDARS measuring $O_3$ concentration, particle concentration, and wind, over a range of about 10 km. They were operated in parallel to generate tomographic observations of the urban boundary layer. The permanent set-up also included an array of 20 T-RH continuous recorders at a 6 m height over the ground, while transect T-RH measurements were occasionally made from the "T-RH Clio" car.

As concerns pollutant distributions in the urban area the measurements were essentially those of the permanent network of the air quality survey service AIRMARAIX : 12 stations within Marseille measuring the concentrations of CO, NO, $NO_2$, $PM_{10}$, hydrocarbon, $O_3$, $SO_2$. In addition, aerosols were measured at pedestrian and roof levels in the city centre (see Chapter 5).

### Main results (ESCOMPTE and UBL/CLU)

All the data obtained during the campaign are being included within the unique ESCOMPTE data base, managed by MediasFrance (for access, contact crob@aero.obs-mip.fr). Most measurements were recorded continuously during the 6 weeks of the campaign, with the exception of the SODAR which were either shut up or operated at reduced power during nights and week-ends. Two types of intensive observation periods (IOP) were more densely documented:

- 5 ESCOMPTE IOPs (for a total of 15 days), generally during breeze situations. During these periods several airplanes measuring the atmospheric composition, turbulence within or at the top of the boundary layer, or wind field transects flew over Marseille to document the urban boundary layer;
- 4 InfraRed days for which airplane equipped with a thermal infrared mapping camera scanned the urban canopy at different times in the day. The influence of

spatial resolution and sensor orientation on the surface temperature measurement were documented by flying in 8 different directions with respect to the sun over the same 3 typical city quarters, especially the city centre around the CAA site monitored by the array of IR radio-thermometers; air temperature at 2 m level was also monitored with the "T-RH Clio" driven under the flight path.

The UBL/CLU data base also includes a set of satellite images, i.e., about 170 images from the NOAA-AVHRR (4 images per day from June 4 to July 16) obtained from the HRPT (Modena, Italy); 2 ASTER high resolution VNIR, SWIR and TIR images covering Marseille (19 June, 5 July) and one the Marignane area (5 June), from the NASA Jet Propulsion Laboratory; the multi-spectral and panchromatic spot images of 17 June.

Finally this data set includes the maps obtained with the specific analysis of data base BDTopo of the IGN (French national geographic institute) which includes such urban objects as buildings, constructions, vegetation, etc. These high resolution maps include urban land uses, roughness parameters, etc.

# 4.4 Other related urban experiments

## 4.4.1 Hamburg

Field measurements were carried out at a 300 m radio transmitter tower located at the edge of a large conurbation (Hamburg, Germany). Using modern instrumentation, high resolution time series of all 3 velocity components were recorded and subsequently analysed. Spectra and integral scales of turbulence were computed and compared with theoretical curves. The results reflect the effects of an about 30 km long urban fetch on the boundary layer formation. They give guidance for the set-up of small-scale boundary layers as they are utilised in physical model studies (Pascheke et al. 2001).

## 4.4.2 Copenhagen

The monitoring of $NO_x/NO$, CO, TSP, $O_3$ etc. has been carried out under the Danish air quality monitoring programme in a street canyon (Jagtvej in Copenhagen and other streets) and at urban background stations during SATURN project. In addition, special measurements of pollutants – especially from traffic - were carried out under the project. The measurements included benzene, toluene and xylenes. In addition, measurements of ultrafine particles were carried out by a Differential Mobility Analyser (DMA) at the street station Jagtvej in Copenhagen, the urban background station at H.C. Ørsted Institute and the street station Albanigade in Odense. The particles were separated in 29 size fractions from 0.01 micron to 0.7 micron. The particle measurements included also $PM_{10}$ (see Chapter 5).

### 4.4.3 St. Petersburg

Since the end of 1998, two DOAS gas analysers have been used to monitor continuously six pollutants, $SO_2$, NO, $NO_2$, $O_3$, benzene and toluene. Simultaneously, meteorological measurements were carried out on the meteorological mast installed on the roof of the building, which hosts the instruments. The monitoring site is located downtown in the street canyon of Pestelya Street with the traffic intensity up to 2000 vehicles per hour. The data collected during two years of observations with the use of DOAS instruments were processed to analyse the air pollution levels on one of the streets in the centre of St. Petersburg. These measurements show that concentrations of most of the species (except $NO_2$) are rather moderate and well inside the Russian ambient air quality standards. These data are directly logged into the municipal automatic system of air pollution monitoring and management, which is used by the city authorities in the decision-making on environmental issues. The results of monitoring and modelling have also been used in preparation of the new version of the national guideline on dispersion modelling finalised in 2001.

### 4.4.4 Tel Aviv

The aim of the study was to develop and test the models for the short term forecasting of air quality, namely $NO_x$ concentration, in the Tel Aviv urban area. Air quality observations were carried out within the urban area at three monitoring sites. Monitoring devices are situated on the roofs of buildings with the inlets for gas measurements at a height of about 15-20 m. The developed statistical models are based on the half-hour averaged measurements of $NO_x$ from five new traffic monitoring stations of the Ministry of Environment located in the central area of the Tel Aviv conurbation, at the street level in proximity to the main arterial roads. The set of input meteorological data for the model development included standard ground-level meteorological parameters and data from diurnal (11:00 GMT) and nocturnal (23:00 GMT) radiosonde launches.

To forecast the morning and evening maximum $NO_x$ concentrations, two multiple-linear regression models have been developed and tested. Using an ordinary least squares method and stepwise regression, the set of predictor variables was obtained (previous day maximum concentration, wind, temperature, relative humidity, cloud cover, etc). The predictant is the maximum $NO_x$ concentration during the morning or evening hours. Test results of one-year predictions (May 2000 – April 2001) versus observed values are presented in Fig. 4.14. The model predicted correctly 96% of the morning occurrences of the admissible $NO_x$ concentrations (less the National Air Quality Standard level of 500 ppb) and 64% of exceeding of the limit level (> 500 ppb). The corresponding results of the evening prediction were 89% and 63%. Similar results were obtained for the period November 2001 – February 2002. Used in combination with the prognostic meteorological data, the developed regression models can be a useful tool for the urban air quality forecasting both in the morning and the evening.

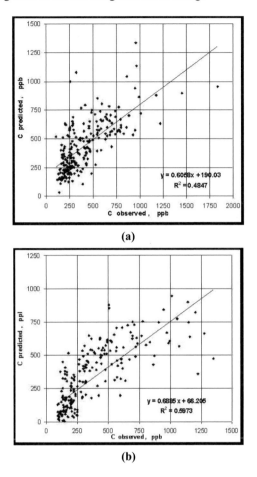

(a)

(b)

**Fig. 4.14.** Maximum NO$_x$ concentrations (May 2000-April 2001) **(a)** morning, **(b)** evening

## 4.5 Summary

During the SATURN project an important effort was performed in order to provide the scientific community with good quality experimental data for local and urban models evaluation. The field campaigns presented in the chapter are summarised in Table 4.5 emphasising the main objectives of the measurements, their duration, major parameters collected during field experiments and the data available for modellers.

**Table 4.4.** Short summary of field campaigns within SATURN

| Scale | Domain | Duration | Objectives | Data available |
|---|---|---|---|---|
| Local scale | Runeberg Str., Helsinki | 1997 | Model evaluation in low wind speed conditions; Studying the chemical transformation processes of $NO_x$ and $O_3$ | Hourly mean concentrations of CO, $NO_x$, $NO_2$, $O_3$ at street and roof levels; Wind speed, wind direction, air temperature, solar radiation |
| | Elimäki | 1995 | Model evaluation Studying the chemical transformation processes of nitrogen oxides and ozone | 3 location, 3 heights on both sides of a road: concentrations of $NO_x$, $NO_2$, $O_3$; traffic densities, relevant meteorological parameters |
| | Nantes '99 | June-July 1999 | Production of turbulent kinetic energy induced by the motion of vehicles; Influence of street surface temperatures on the pollutant dispersion; Model validation used for studying the dynamic and thermodynamic structure of the urban canopy layer | Concentrations of CO, $NO_x$, $SO_2$, dust, $O_3$ with 15 minutes resolution at different levels. Radiation budget components, traffic counts, air temperature and temperature of walls, wind direction, wind speed, fluctuations of the three wind components |
| | Klagenfurt | 4 month | Dispersion conditions near tunnel portals Model development and application | $SF_6$ with 30 minutes resolution, meteorological data. |
| | Podbielski Str. | | Validation of microscale dispersion models | Wind speed and direction, traffic, benzene, soot, $NO_2$, NO, CO |
| | Göttinger Str. | August 2001 | Validation of microscale dispersion models | Continuous measurements of air pollutants and meteorological parameters in 12 locations at street and roof levels |
| | London | Summer '99 Winter '00 | Road user exposure to $PM_{2.5}$ | Road user exposure to $PM_{2.5}$ for three modes of transport bicycle, bus and car |
| Urban scale | Lisbon Region-LisbEx (200×200 km²) | July 1996, July 1997 | Characterisation of meteorology and air quality during typical summer conditions; Breeze circulation in coastal zone; Atmospheric boundary layer vertical structure Model evaluation | Surface measurements of CO, $NO_x$, $O_3$, $SO_2$ : temperature, relative humidity, wind speed and direction; Vertical structure up to 8000 m for pressure, temperature, wind speed, wind direction and $O_3$ |

**Table 4.4.** (cont.)

| Monitoring type | | | | |
|---|---|---|---|---|
| | Graz (26×34 km²) | Winter 98/99, Summer 99 | Dispersion conditions of cities in valleys; Influence of local wind system and strong temperature inversion on air quality; Vertical structure of the boundary layer; Model validation; | Concentrations of NO, NO₂, SO₂, TSP, CO, hourly vertical O₃ profiles; Meteorological parameters from monitoring stations; hourly vertical profiles of wind speed and direction, temperature, humidity at four locations |
| | Milan | 7 campaigns during 1997-2001 | Chemical regimes near and inside the city plume; Vertical chemical characterisation; Urban aerosol formation; | Ground level and vertical profiles (0-1000m) of wind, turbulence and temperature. Concentrations of NMHC, C2-C9, HC, BTEX, CH₄, aldehydes, carbonyl compounds, PM₁₀, PM₂.₅, CO, NOₓ, O₃, H₂S, SO₂, HNO₃, HNO₂; Vertical profile of O₃ and aldehydes |
| | Marseille ESCOMPTE, UBL/CLU (100×100 km²) | June-July 2001 | Meteorological and chemical conditions prevailing during photochemical episodes Four-dimensional structure of Urban Boundary Layer Database to test urban energy exchange and high resolution meteorological and chemistry-transport models | Concentrations of O₃, NOₓ, VOCs, PM₁₀, SO₂; emission/deposition velocities of trace gases (O₃, NOₓ), vertical thermodynamic and O₃ profiles, Meteorological basic parameters (wind, temperature, moisture and radiation), surface energy budget, turbulent and radiation fluxes at different heights within urban canopy level, heat exchanges between building inside and outside |
| | Hamburg | | Provide guidance for the set-up of small-scale boundary layers used in physical models | 3 wind velocity components at 300 m height |
| | Copenhagen | | Air quality in street canyons | NOₓ/NO, CO, TSP, O₃, benzene, toluene, xylenes, PM₁₀, ultrafine particles of 29 size fractions |
| | St. Petersburg | Since 1998 | Guidance for the city authorities in decision making on environmental issues | SO₂, NO, NO₂, O₃, benzene, toluene; meteorological parameters |
| | Tel Aviv | 1999-2002 | Diurnal, weekly and season variation of pollutants, Short-term forecast | NOₓ, NO₂, O₃ |

# Chapter 5: Particulate Matter in Urban Air

J. Kukkonen[1], L. Bozó[2], F. Palmgren[3], R.S. Sokhi[4]

[1] Finnish Meteorological Institute, SF-00880 Helsinki, Sahaajankatu 20E, Finland

[2] Hungarian Meteorological Service, H-1675 Budapest, P.O.Box 39, Hungary

[3] National Environmental Research Institute, Frederiksborgvej 399, DK-4000 Roskilde, P.O.Box 358, Denmark

[4] Atmospheric Science Research Group, Science and Technology Research and Innovation Centre, University of Hertfordshire, College Lane, Hatfield, Herts, AL10 9AB, UK

## 5.1 Introduction

Research within SATURN has focussed on improving our understanding of the physical and chemical properties of particulate matter, and our capability for modelling particulate matter concentrations in urban areas. Campaigns have been conducted at various urban centres on $PM_{10}$, $PM_{2.5}$ and on finer size fractionated samples. Measurements have been made to reveal the chemical composition, and particle number and mass concentrations at various urban locations. Models have been developed, evaluated and applied in order to predict aerosol processes, and the number and mass concentrations, of the coarse and fine fractions in urban areas.

### 5.1.1 The reasons for investigating urban particulate matter

Studies of long-term exposure to air pollution, especially to particulate matter (PM), suggest an increased mortality, increased risk of chronic respiratory illness, and of developing various types of cancer. World Health Organisation (WHO

1999) has estimated that in Europe air pollution has caused 168.000 (range of estimate 100.000 – 400.000) excess deaths annually; in the United States the corresponding figure has been estimated to be approximately 100.000. The best estimate on the reduction in life expectancy in Central Europe is about 1 year.

Fine PM particularly may be causing a significant burden of disease and excess deaths in Europe and North America. However, it is not known, which chemical and physical characteristics of the PM are responsible for these effects, and which source categories are responsible for the most harmful exposures.

The chemical and physical properties of PM are important for assessing environmental impact and deposition in the lungs as well as adverse health effects (Harrison et al. 1999; Morawska et al. 1999; Kleeman et al. 2000). In addition to size and number concentration, important properties include state (liquid/ solid), volatility, hygroscopicity, chemical composition (content of organics, metals, salts, acids etc.), morphology and density. These properties also need to be taken into account in selecting methods for PM emission regulation and control. A reliable assessment of urban PM pollution and the subsequent adverse health effects is therefore crucial in terms of the promotion of public health (WHO 2000).

### 5.1.2 Review of experimental field campaigns on particulate matter in SATURN

Several experimental campaigns were carried out within the framework of SATURN during the past years. Different European urban areas were investigated in Denmark, Finland, France, Greece, Hungary, Sweden and the United Kingdom. Field studies aimed mainly at investigating particle size distributions and their chemical properties. Key objective of the studies was to conduct sufficiently detailed measurements in order to enable source apportionment investigations and characterisation of particles, including the fine and ultrafine particle fractions.

Measurement campaigns of particulates were carried out in Copenhagen at streets, urban background and regional background in order to characterise urban PM, including size distributions as well as chemical composition and physical properties. Special measurements of ultra-fine particles were carried out in order to estimate the emission and number size distribution of the actual car fleet, separately for diesel and petrol vehicles (e.g., Wåhlin et al. 2001a).

In the Helsinki metropolitan area, virtual and Berner low-pressure impactors were applied to collect aerosol samples. Size distribution and chemical composition of the particles were analysed (e.g., Pakkanen et al. 2001a, b, c). In Budapest, source profiles of toxic elements were determined and the source-receptor relationships were analysed by the chemical mass balance method (e.g., Bozó et al. 2001).

A field measurement campaign was performed in urban air and in a road tunnel in Stockholm. The tunnel experiment has provided source profiles and emission factors for gasoline and diesel vehicles. Both particulate species such as metals, polycyclic aromatic hydrocarbons (PAH), elemental and organic carbon and gases (NO, $NO_2$, CO and several volatile hydrocarbons) were measured in the tunnel (e.g., Kristensson et al. 2001). The same set of compounds was also measured in an urban ambient air campaign.

Sampling campaigns were carried out in Hatfield, UK by the University of Hertfordshire in the summer of 2000 and spring 2001. Gravimetric measurements at various size fractions were coupled with detailed elemental analysis. The aim of their research was to better understand the mass and chemical distributions of size fractionated aerosols. Sampling for these campaigns was conducted at a semi-urban roof top site (Sokhi & 1999 and & 2001).

Research at Imperial College on source apportionment has used modelling of $PM_{10}$ concentrations across London to differentiate contributions from local major roads, emissions from central, inner and outer London, imported secondary particulate contributions from longer range transport, and the coarser fraction between 2.5 and 10 μm. This has been used to assess the benefits of a range of potential future scenarios involving both technological and traffic reduction measures.

## 5.1.3 Review of modelling particulate matter in SATURN

Mathematical models have proved to be very useful for a range of air quality applications. Examples of these and, in particular those related to particulate matter, can be found in Sokhi et al. (2000b) and Sokhi and Bartzis (2002). Within the framework of SATURN, a range of aerosol process, emission and dispersion models have also been developed in order to evaluate urban PM concentrations.

The Danish street pollution model and the urban background model were refined to include treatment of particulate matter. The application of inverse modelling has led to estimates of emission factors of particles from the Danish car fleet under real life driving conditions. The emission factors comprising size distributions and particle composition are important as input to air quality and exposure models.

In Finland, the aerosol dynamical model MONO32 was applied; this model has been originally developed at the University of Helsinki by Pirjola and Kulmala (2000). The model takes into account gas-phase chemistry and aerosol dynamics, i.e., the processes of nucleation, coagulation, condensation, evaporation and deposition. The particles are classified into four different size modes, which are assumed to be monodispersive. The first objective was to evaluate quantitatively the impact of various chemistry and aerosol processes on aerosol evolution (Pohjola et al. 2003).

Based on road tunnel measurements of particle size distributions in Stockholm, an aerosol dynamical module has been developed to study the dynamics of different particle sizes in traffic tunnels and other heavily trafficked urban environments, such as, e.g., street canyons (Gidhagen et al. 2002). The road tunnel represents a well-controlled environment in order to produce good-quality data for model evaluation. In addition, tunnel data may be used to determine vehicle emission factors and to investigate aerosol dynamics and dispersion.

A fairly simple model was developed for predicting the concentrations of $PM_{2.5}$ in urban areas (Tiitta et al. 2002). This model also includes a method for evaluating the regionally and long-range transported fine particle fraction (Karppinen et al. 2002). The model was tested against the results from a measurement campaign in a suburban environment near a major road in Kuopio, Central Finland. The mass concentrations of fine particles ($PM_{2.5}$) were measured simultaneously at four distances from a major road, together with traffic flows and relevant meteorological parameters.

The US EPA Gaussian plume (CALINE 4) and ADMS-Urban models were employed to calculate $PM_{10}$ concentrations near to a major motorway (M25) encircling the City of London. The model was used to calculate 24 hour and annual means of $PM_{10}$; a comparison with measurements for 1996 showed a close agreement (Sokhi et al. 1998 and 2000a). The study showed that roughly half of the modelled PM10 concentrations could be attributed to the road traffic contribution.

A semiempirical model has been developed for evaluating the $PM_{10}$ concentrations in urban areas (Kukkonen et al. 2001a). The basic model assumption is that local vehicular traffic is responsible for a substantial fraction of the street-level concentrations of both $PM_{10}$ and $NO_x$, either due to primary emissions or vehicle driven material from street surfaces. The model performance was evaluated against the measured $PM_{10}$ data from five air quality stations in the Helsinki area in 1999.

### 5.1.4 Review of investigations on exposure to particulate matter in SATURN

The exposure of road-users to $PM_{2.5}$ was studied both experimentally and by modelling in London (e.g., Adams et al. 2001a, b). One of the main results was that people moving along city streets are exposed to levels of pollution that are not only substantially higher than urban background fixed-point measurements, but also higher than roadside and kerbside measurements. In Helsinki, a detailed population exposure model was developed that can be utilised for evaluation of the whole urban population to PM (Kousa et al. 2002).

## 5.2 Sources and emissions of particulate matter

### 5.2.1 Review of various emission sources of urban particulate matter

Road traffic is a major source of particles in urban air in most European cities. Traffic related particulate matter is not only emitted directly from the vehicle exhaust but also includes particles from wear on road, tires and brakes, and airborne dust from road surfaces.

As an example, one can consider the contributions to measured concentrations of $PM_{2.5}$ at a roadside measurement location originating from various source categories. The total measured concentration of $PM_{2.5}$ can be written as (modified from Tiitta et al. 2001; Kukkonen et al. 2001a):

$$PM_{2.5} = PM_{2.5}^{tr,e} + PM_{2.5}^{tr,n-e} + PM_{2.5}^{st} + PM_{2.5}^{bg,urb} + PM_{2.5}^{bg,lrt} + PM_{2.5}^{wind} \qquad (5.1)$$

where the superscripts 'tr,e' and 'tr,n-e' refer to the primary (exhaust) and non-exhaust contributions of vehicular traffic from the nearest roads and streets, respectively. The superscript 'st' refers to stationary sources, and the superscripts 'bg,urb' and 'bg,lrt' refer to the urban, and regionally and long-range transported (LRT) background, respectively. The superscript 'wind' refers to wind-driven PM from various surfaces (excluding non-exhaust vehicular traffic emissions). Equation (5.1) is useful for illustrating the multiple source types of urban airborne PM. Clearly, this is not the only option in order to categorise various emission sources.

### 5.2.2 The relative importance of various particulate matter emission sources

The particles may grow by coagulation of primary particles, and by condensation of gases on particles. The fine particles (accumulation mode in the range from 0.1 to 2 μm) are typically formed by chemical reactions (e.g., $SO_2$ and $NO_x$ to form sulphate and nitrate), or other relatively slow processes in the atmosphere; the fine PM are therefore commonly aged particles. The coarse particle mode can be defined as the particles with diameters larger than 2.5 μm, which in urban areas are formed typically mechanically by abrasion of road material, tyres and brake linings, construction works, soil dust raised by wind and traffic turbulence.

In the UK, there are numerous sources of fine particles that are recognised in urban centres (QUARG 96, APEG 1999); these include contributions from natural sources, road transport, stationary combustion and industrial processes. Road transport nationally can contribute about 25 % of $PM_{10}$, whereas in urban centres the contribution rises to approximately 80-90 % (QUARG 96, APEG 1999). The coarse fraction (particles with aerodynamic diameters from 2.5 to 10 μm) tend to consist mainly of soil, sea salt, biogenic and airborne dust components, whereas the fine fraction (diameters below 2.5 μm) consists of components from road vehicles (mainly diesel) and stationary combustion processes, and secondary aerosols (such as sulphates, nitrates and ammonium) as well as transboundary contributions.

In Nordic countries, atmospheric long-range transport constitutes an important part of the total urban background $PM_{2.5}$ concentration (e.g., Johansson et al. 1999; Pakkanen et al. 2001b, c; Karppinen et al. 2002). In urban areas in Sweden, it even constitutes the most important single contribution to both urban background $PM_{2.5}$ and $PM_{10}$, which is clearly illustrated by the spatially uniform urban background concentrations of these particulate fractions (Areskoug et al. 2000).

The long-range transported aerosol dominates the accumulation mode of the aerosol size distribution even in urban areas, whereas its contribution to the ultrafine and coarse mode is small, in comparison to that of the local particle emissions. For cities in Nordic countries, the coarse particle fraction is important for the elevated $PM_{10}$ concentrations, especially in spring (e.g., Johansson 1999; Pohjola et al. 2000 and 2002).

Like in other urban areas, in Budapest, Hungary, the highest $PM_{10}$ concentrations tend to occur in the vicinity of major urban roads (Bozó et al. 2001), indicating that local traffic is mainly responsible for the elevated coarse particle concentrations. The amount of major industrial pollution sources has substantially decreased in urban areas in Central Eastern Europe during the last decade. In Budapest, heat energy production is based mainly on natural gas, regarding both thermal power plants and domestic heating.

As stated earlier, a major contribution to particulate pollution in urban areas is believed to be attributed especially to emissions from diesel-powered vehicles (Palmgren & 2001). Ultra-fine particles emitted from petrol as well as diesel engines are formed at high temperature in the engines, in the exhaust pipe, or immediately after emission to the atmosphere. Some of these particles may be in the so-called nucleation mode (nanoparticles < 50 nm). The dominating ultrafine particle mode has a number concentration peak in the range of 20-50 nm. The ultrafine particles in urban air are mainly emitted from diesel and petrol fuelled vehicles.

In Copenhagen, hourly elemental and organic carbon (EC/OC) measurements of $PM_{10}$ were carried out in a busy street, related to measurements of other traffic-originated pollutants. The statistical correlations between EC and CO and also OC and $NO_x$ /CO were relatively low. However, a clear correlation between EC and $NO_x$ was observed, indicating a significant contribution to EC from diesel traffic (Palmgren et al. 2001).

Measurement campaigns carried out in two cities in Denmark included monitoring of the ultrafine particles from traffic under normal driving conditions and in ambient air, in order to be able to establish the relationship between the sources and the exposure of the population. Measurements of ultrafine particles were carried out by a SMPS (Scanning Mobility Particle Sizer) with a high time resolution that corresponds to the variation in traffic and meteorological variables.

The particles were separated in 29 size fractions, ranging from 0.01 to 0.7 μm. It was shown that particles originating from diesel engines include a large nanoparticle fraction (< 30 nm). These very small particles are probably mainly droplets, which are formed in or immediately after the tailpipe. The sulphur content plays a key role as condensation nuclei for condensation of fuel, lubrication oil and volatile combustion products. The particle number concentration may therefore depend strongly on the sulphur content of the diesel oil consumed (Fig. 5.1).

Particles from diesel as well as petrol engines comprise also solid, mainly carbonaceous, particles, which may have a crucial role regarding the adverse health effects. The measurements of ultrafine particles in Copenhagen have indicated this mode of the particle size distribution (Wåhlin et al. 2001b).

The results from Stockholm (Johansson *&* 2001) and Nantes (Despiau *&* 2001)suggest that local traffic is the main source of ultra-fine particles, and most of the particles close to traffic are in the range from 3 to 30 nm. The size distributions change rapidly with distance from the traffic sources. At the rooftop location, the size distributions shift towards larger sizes compared with those in the street level.

In residential areas of North European cities, biomass burning for domestic heating may be an important source of fine particles during winter. For instance in Sweden wood is most frequently burned in boilers constructed for multiple energy sources (oil, wood and electricity). Recently, low emission boilers have been introduced into the market, but stoves have come into use as an additional heating device, commonly also in urban areas.

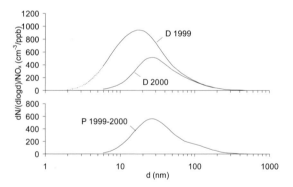

**Fig. 5.1.** The number size distributions in the winters 1999 and 2000, respectively, at Jagtvej in Copenhagen before and after the reduction of the sulphur content in diesel fuel in Denmark in July 1999 (Wåhlin et al. 2001b). The upper graphs are diesel distributions, which showed a pronounced shift (probably mostly due to diesel cars with oxidising converters), and the lower graph is a distribution which did not change (probably due to a combination of petrol and diesel traffic)

Emissions from wood combustion may be important for PM, but also for many particle-bound hydrocarbons, such as polycyclic aromatic hydrocarbons. The aerosol originated from biomass combustion may substantially differ from that originating from vehicle exhausts, both regarding chemical composition and particle size distribution. For instance, PM from biomass burning may contain a larger fraction of hygroscopic particles (Hedberg et al. 2002).

### 5.2.3 Evaluation of the vehicular emissions of particulate matter

The assessment of the emission of pollutants is normally based on dynamometer studies with different driving cycles for a few vehicles. Dynamometer cycles are essential in order to establish uniform emission standards for regulatory purposes and for testing of new technologies. However, such measurements do not reflect the real driving conditions and the level of maintenance of the actual vehicle fleet.

The control parameter is normally the mass of the particles that will be dominated by the coarse particulate fraction. The mass includes often only the solid part of the particles, which are measured after heating of the exhaust gas to, e.g., 300°C; this will remove semi-volatile compounds, e.g., organic condensates, which could nevertheless be present in the urban air. There is therefore a need for on-road emission estimates of air pollutants from the actual fleet.

The emission factors for particle distributions from different vehicle categories can also be estimated by inverse modelling. These estimates are based on time series of particle spectra from streets and urban background locations, as well as detailed traffic data and local meteorology at rooftop level.

The model CAR-FMI (e.g., Kukkonen et al. 2001b) was extended to allow for the emissions and dry deposition of PM. The vehicular $PM_{2.5}$ emissions were modelled to be dependent on vehicle travel velocity (ranging from 0 to 120 km h$^{-1}$), separately for the main vehicle categories. Light-duty vehicles were classified into three categories: (i) gasoline-powered cars and vans without a catalytic converter; (ii) gasoline-powered cars and vans equipped with a catalytic converter; and (iii) diesel-powered cars and vans. Similarly, heavy-duty vehicles were classified into four categories: (i) diesel-powered trucks with a trailer; (ii) diesel-powered trucks without a trailer; (iii) diesel-powered buses; and (iv) buses powered by natural gas.

Numerical correlations of PM emissions were fitted in terms of vehicle travel velocity, separately for each of the above-mentioned seven vehicle categories. These correlations are based on nationally conducted vehicle emission measurements (Laurikko 1998).

# 5.3 Spatial and temporal variation – temporal trends

## 5.3.1 Spatial and temporal variation of particulate matter in urban areas

To estimate the extent of impact at various urban centres, we need to understand the temporal and spatial variations of PM concentrations.

The main objective of the UK study was to investigate the temporal variation of $PM_{10}$ and $PM_{2.5}$ concentrations at two London sites. Data includes measured PM concentrations ($PM_{10}$ and $PM_{2.5}$) from Marylebone Road (roadside site) and Bloomsbury (urban centre site) for the period 1997-1999. Data was obtained from 'The UK National Air Quality Archive'.

Monthly variation of particulate concentrations at the two sites has been investigated, and correlation with other air quality and selected meteorological parameters has been analysed. Fig. 5.2 shows monthly variations of $PM_{10}$ and $PM_{2.5}$ at these two sites. The correlation analysis suggests that $PM_{10}$ and $PM_{2.5}$ have the highest correlation amongst each other and a highly significant correlation with other air quality parameters, such as CO and $NO_x$. This obviously confirms road traffic as one of the main sources. On average the PM concentrations were found to be about 40 % higher at the roadside site compared to the urban centre site, confirming the importance of particles from surrounding areas.

These two sites also show a moderate monthly variation for both years, except for the coarse fraction at the roadside site (1999), which shows a peak during the late summer and early autumn months. The peak is caused by high hourly concentrations (> 50 $\mu g/m^3$) of $PM_{10}$ which were observed at the roadside site for about 10% of the time. A possible reason for the high coarse fraction concentrations could be the occurrence of generally dry conditions coupled with wind speed of around 4-5 m/s increasing the dust content in the local atmosphere. The coarse fraction is higher at the roadside site compared to the background site, indicating the importance of vehicle turbulence as a possible mechanism for in introducing this fraction into the local atmosphere.

Seasonal variation of the $PM_{10}$ and $PM_{2.5}$ concentrations at some stations in the Helsinki Metropolitan Area during one selected year (1998) are presented in Fig. 5.3 (Pohjola et al. 2000, see also Pohjola et al. 2002).

The seasonal variation of the $PM_{10}$ concentrations in Fig. 5.3 is very pronounced. The $PM_{10}$ concentrations have maximum values in spring at all measurement stations; these are mostly caused by vehicle induced airborne material from street surfaces. The fine particulate matter ($PM_{2.5}$) concentrations also show a maximum in spring that may be caused by vehicle induced airborne material or unfavourable meteorological conditions, or both reasons. However, there is a substantial year-to-year variation; for instance during the previous year (1997), there was no clear seasonal variation of the monthly averaged $PM_{2.5}$ concentrations.

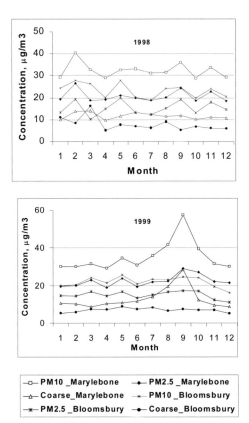

Fig.5.2. Monthly variation of PM$_{10}$ and PM$_{2.5}$ at two London sites

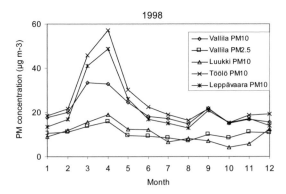

**Fig. 5.3.** Seasonal variation of PM$_{10}$ concentrations (monthly averages) in Helsinki, at the stations in Vallila and Töölö (urban, traffic sites) and Leppävaara (suburban site), together with PM$_{2.5}$ concentrations at Vallila and regional background PM$_{10}$ concentrations at Luukki (Pohjola et al. 2000)

In Stockholm, the highest PM$_{10}$ levels are also observed during spring from March to April, caused mainly by vehicle induced airborne material from street surfaces. Most exceedances of the daily mean EU limit value for PM$_{10}$ occur during the spring (Fig. 5.4). The spring maximum is particularly pronounced at the kerb sites close to traffic, but can also be seen at urban background sites. For fine particulate matter fraction, PM$_{2.5}$, there is almost no annual variation and the urban background concentration is almost the same as the rural background.

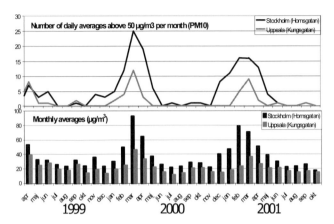

**Fig. 5.4.** Upper panel: Number of days with average PM$_{10}$ concentration higher than 50 µg/m$^3$ (EU limit) in Stockholm (upper line) & Uppsala (lower line). Lower panel: Monthly average PM$_{10}$ concentration at the same sites (Stockholm: black bars, Uppsala: grey bars)

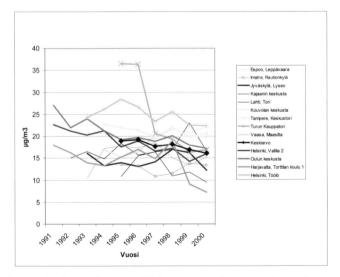

**Fig. 5.5.** The trends of annual average PM$_{10}$ concentrations at various urban measurement stations located in 12 cities in Finland during the 1990's. The average of all stations is denoted as bold black line with squares (Pietarila et al. 2001)

### 5.3.2 Temporal trends of particulate matter in urban areas

The trends of annual average $PM_{10}$ concentrations are presented in Fig. 5.5 for various urban measurement stations in Finland since 1991. The data includes urban measurement stations located in 12 cities. These trends show a substantial variability both from one station to another, and from year to year. However, the annual average value of all stations shows a slight decreasing trend. The measurement data of fine particulate matter, $PM_{2.5}$ is not yet sufficient for detailed analysis to reveal any reliable year-to-year trends.

Since the measurements started in 1994, the $PM_{10}$ concentrations in Stockholm (Sweden) have shown no clear temporal trend. Urban background concentrations of $PM_{10}$ in Stockholm range from 13.5 to 16 $\mu g/m^3$. While local vehicle exhaust emissions have probably significantly decreased during this period, due to the renewal of vehicle fleet and the introduction of cleaner fuels, they do not substantially contribute to the total $PM_{10}$ concentrations. The main sources of $PM_{10}$ in Stockholm are background air transported to the urban area and vehicle induced airborne material from street surfaces.

## 5.4 Field campaigns involving particulate matter

In SATURN, field measurement campaigns were conducted in Denmark, Finland, France, Greece, Hungary, Sweden and the United Kingdom. In the following, five of the above mentioned campaigns are described in a structured format that includes (i) major scientific issues to be investigated with the campaign, (ii) explicit relevance to SATURN aims, (iii) experimental set-up and (iv) main results.

### 5.4.1 Field measurement campaigns in Copenhagen and other cities in Denmark

*Major scientific issues investigated (Copenhagen)*

The objective was to measure and determine the particle emissions under normal driving conditions in ambient air, in order to establish the relationship between the sources and the exposure of the population (Palmgren & 2001).

The population spends most of the time indoors. The ambient (outdoor) air quality has a significant effect on indoor pollution levels (Lai and Nazaroff 2000). There is thus a strong need for quantifying the processes governing the particle composition and size distributions also in the indoor environment.

### Relevance to SATURN aims (Copenhagen)

A four-year project on particle studies was initiated in 2000/2001. The main objectives are (i) to characterise the geographic and temporal variability in particle composition and size distributions in Danish ambient air, (ii) to determine particle emission factors for various vehicle categories, (iii) to determine indoor - outdoor relationships for buildings in busy streets and (iv) to develop the Danish air quality models for local (OSPM), urban (BUM) and regional scales to include particles and to validate the models for particles.

### Experimental set-up (Copenhagen)

Most of the measurements in these studies were performed in central Copenhagen in a street canyon, Jagtvej, which is a 10 m wide main road and during rush hours in practise a 4 lane road. Both sides of the roadway have bicycle lanes and pavement. In addition, Jagtvej is lined on both sides by 5-6 storey buildings. The traffic density is approx. 26,000 vehicles per 24 hours, including 6-8 % heavy vehicles, i.e. buses, lorries and larger vans.

A fixed monitoring station of the Danish Air Quality Monitoring Programme has been in operation at this location for many years (Kemp and Palmgren 2000). Data from this station include half-hour measurements of $NO_x$ (sum of the nitrogen oxides NO and $NO_2$), CO (carbon monoxide), and other traditional pollutants. In addition, 24 hours particle filter samples were collected of TSP (Total Suspended Particulates) and $PM_{10}$. Other measurement campaigns were performed at H.C. Andersen's Boulevard (approx. 60,000 vehicles per day), a street in Copenhagen. Some measurement campaigns were performed at Albanigade (22,000 vehicles per day, 12 % diesel), a street in Odense, a city of 180,000 inhabitants about 150 km west of Copenhagen. All three stations are included in the Danish Air Quality Monitoring programme with the above mentioned measurement programme. The urban background air pollution was measured at rooftop stations in the Danish Air Quality Monitoring programme.

A Scanning Mobility Particle Sizer (SMPS) was used to measure the fine and ultrafine particles. Campaigns of $PM_{10}$ measurements with a time resolution of half an hour were carried out at Jagtvej in Copenhagen by Tapered Element Oscillating Microbalance. The measurements of the $PM_{10}$ elemental and organic carbon (EC/OC) were performed using the automatic speciation analyser 5400 Ambient Carbon Particulate Monitor (R&P).

Studies of exchange of particles in outdoor/indoor air and the transformation took place in the street canyon, Jagtvej, in Copenhagen and in an apartment along the street. Four measurement campaigns were performed in 2001/2002.

## Main results (Copenhagen)

The Constrained Physical Receptor Model (COPREM) has been used for the source apportionment. For the apportionment it is important to find the emission profiles for the sources. In Wåhlin et al. (2001a), it has been shown that the emission profiles, in principle, can be determined, if the average $CO/NO_x$ emission ratios for the two traffic categories are known. As the $CO/NO_x$ emission ratio for diesel is practically zero compared with the ratio for petrol, the critical unknown parameter is the average $CO/NO_x$ emission ratio for petrol.

This ratio was determined on the basis of traffic counts in Albanigade, Odense, and a specific solution was found for the particle concentrations. The solution for the total particle concentrations in an average week at Albanigade is shown in Fig. 5.6. The small diurnal variation of the non-traffic contribution is different from the traffic pattern and is probably due to the general activity in the city (Wåhlin et al. 2001a).

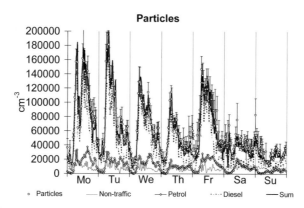

**Fig. 5.6.** The average weekly cycle of particle number concentrations measured at Albanigade, 3/5/99-20/5/99, and the fitted contributions from the different sources (Wåhlin et al. 2001a)

## 5.4.2 A field measurement campaign in Helsinki, Finland

### Major scientific issues investigated (Helsinki)

The principal aim was to reveal the influence of long-range transport and major local sources on the concentrations of various chemical species in ultrafine ($PM_{0.1}$), fine ($PM_{2.3}$) and coarse ($PM_{2.3-15}$) aerosol particles. In order to obtain reliable results, the performance of different size-segregating aerosol collectors was compared for 14 different ions. Gas-particle interactions and particle deposition characteristics were also studied.

### Relevance to SATURN aims (Helsinki)

The results are useful in estimating the health effects of atmospheric particles and in planning emission abatement strategies. Further, the results can be used as input data for modelling studies, and contribute to an improved understanding of urban particulates and source-receptor relationships.

### Experimental set-up (Helsinki)

In April 1996 - June 1997, size-segregated atmospheric aerosols were measured simultaneously at Vallila, an urban site in Helsinki, and at Luukki, a rural site in Espoo. At both sites, separated by about 20 km, virtual impactors (VI, 53 samples) and Berner low-pressure impactors (BLPI, 10 samples) were operated in parallel. In addition, black carbon (BC, particle diameter $< 2.5$ μm) was monitored at the urban site, and gaseous $SO_2$, $NO_x$ and $O_3$ were monitored at both sites.

The different size fractions were analysed utilising inductively coupled plasma - mass spectrometry (ICP-MS) and ion chromatography (IC). About 100 samples were analysed using instrumental neutron activation analysis (INAA). Quality control was introduced by comparing the parallel sampling and parallel analysis (Pakkanen et al. 2001a; Pakkanen and Hillamo 2002).

### Main results (Helsinki)

Average fine particle ($PM_{2.3}$) mass concentrations were 11.8 μg/m$^3$ and 8.4 μg/m3 and those of coarse particles ($PM_{2.3-15}$) were 12.8 μg/m$^3$ and about 5 μg/m$^3$ at the urban and rural sites, respectively, indicating that relatively strong local sources existed around the urban site. On average, fine particle mass at the urban site consisted of not-analysed material (43 %, mainly carbonaceous material and water), sulphate (21 %), nitrate (12 %), crustal material (12 %), ammonium (9 %) and sea-salt (3 %), while coarse particle mass consisted of crustal material (59 %), not-analysed material (28 %, mainly carbonaceous compounds and water), sea-salt (7 %), nitrate (4 %) and sulphate (2 %) (Pakkanen et al. 2001b). According to an earlier study (Pakkanen et al. 2000) made at the urban site, the average black carbon (BC) concentration in fine particles was 1.4 μg/m$^3$, of which 0.4 μg/m$^3$ was long-range transported. On average, local traffic contributed to the BC concentrations 63 % on working days and 44 % on Sundays.

Average mass concentrations of ultrafine particles ($PM_{0.1}$) were about 0.5 μg/m3 at both the urban and rural sites (Pakkanen et al. 2001a). The ultrafine mass seemed to consist mainly of carbonaceous material (70 %) and water (10 %), since the average concentrations of analysed components were low (at the urban site sulphate 0.032 μg/m$^3$, ammonium 0.022 μg/m$^3$, $Ca_2^+$ 0.005 μg/m$^3$ and nitrate 0.004

0.004 µg/m$^3$). The analyses of ultrafine particles consisted of more than 40 chemical components, whose average concentration was above the limit of detection.

Detailed mass size distributions (aerodynamic particle diameter 0.03 - 15 µm) of particulate mass and 27 elements were utilised in estimating fine particle contributions arising from local sources versus long-range transport (Pakkanen et al. 2001c). It was estimated that at the urban site 46 % of fine particle mass and most of, for instance, fine particle Ni (nearly 100 %), Cu (77 %), V (72 %), Bi (67 %), and Sb (55 %) was of local origin. The local contributions of Tl (23 %), As (26 %), Pb (31 %) and Cd (33 %) were estimated to be low. Local road dust was the dominant source for fine particle crustal elements, such as, Al, Ba, Ca, Fe, Mg and Ti, leading to local contributions close to 100 % for these elements. The average particle mass size distributions measured at the urban and rural sites are presented in Fig. 5.7.

Concentrations and size distributions of particle bound low-molecular-weight dicarboxylates were studied at the urban and rural sites (Kerminen et al. 2000). Compared to winter, concentrations were much higher during summer, the likely principal sources being (i) secondary production in the long-range transported air masses and (ii) local traffic.

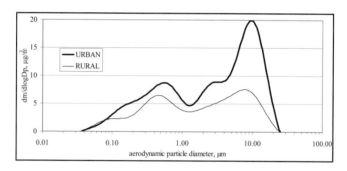

**Fig. 5.7.** Average particle mass size distributions at the urban and rural site in the Helsinki Metropolitan Area

### 5.4.3 A field measurement campaign in Budapest, Hungary

#### *Major scientific issues investigated (Budapest)*

The origin of aerosol particles is of crucial importance for environmental management in an urban scale. Source characterisation and receptor modelling applied in connection with the measurement campaign provides quantitative estimates of the source contributions on ambient air PM. In contrast to dispersion modelling, no highly detailed meteorological and emission data is needed.

As stated previously, aerosol particles are of great concern due to their adverse health effects. In order to better formulate mitigation strategies to reduce such impacts in urban areas, an important first step is to identify the major sources and composition of ambient particles. Previously, regional scale investigations have been performed in Hungary regarding the transport and chemical composition of particles (Bozó 2000; Havasi et al. 2001). Fine aerosol sampling and related source-receptor modelling for Budapest was started in 1997.

## Relevance to SATURN aims (Budapest)

Local sampling and measurements, as well as the Chemical Mass Balance method has been applied to estimate the source contributions to ambient air concentration levels of trace elements in Budapest. The general aim is to optimise the emission control strategies that will be effective in reducing the ambient concentrations of pollutants considered. Effects of regional pollution could also be pinpointed during the sampling and measurements campaign.

The experimental results are utilised in combination with the ADMS model computations (these are described in more detail in the modelling part of this section) in order to locate the most polluted areas in Budapest. The temporal variation of particle concentrations can also be evaluated based on measurements conducted during several years.

## Experimental set-up (Budapest)

Source profiles of Cd, Cu, Ni, Pb, V and Zn for waste incineration, traffic, oil and coal burning were applied for model calculations. Aerosol sampling for fine size range aerosol particles was carried out by Harvard-type impactors at the sources and at two receptor points, in the downtown of Budapest during November and December in 1999, and summer months in 2000 and 2001.

One of the sampling sites (OKI) is located at the SE edge of the downtown area with relatively heavy traffic in its vicinity, while another site (ELTE) is located in the vicinity of the Danube River that crosses the city in the N-S direction. This site is fairly intensively ventilated; it can therefore be expected that a mixture of various source categories be detected at that site.

## Main results (Budapest)

It was found that substantial amounts of Zn, Pb and Cu are originated from a waste incinerator that is located in Budapest. Regarding the traffic profile, the most important element is still Pb; however, its relative contribution has decreased rapidly during the past five years.

It was concluded that waste incineration provides the most significant contribution to the toxic metal load in Budapest (from 65 to 70 %). The relative contribution of traffic sources ranges from 11 to 17 % (Table 5.1). Coal burning has no significant importance in Budapest, regarding the receptor profile. The coal consumption has significantly decreased in Budapest during the past decade, since it has been replaced by natural gas at most industrial, power-production and residential sources.

In co-operation with NOAA, three-dimensional backward trajectories were also computed in order to estimate the origin of air masses. Episodes with elevated concentrations of Ni and V measured in Budapest are accompanied by Southerly winds, indicating the effects of an oil refinery located at a distance of 25 km South of Budapest.

Measurement campaigns are continued at three receptor points in Budapest as well as in the vicinity of relevant stationary and mobile sources. The objective is to extend the investigations towards the relationships between meteorological conditions and receptor profiles.

**Table 5.1.** Relative contribution (%) of source categories to ambient trace element concentrations (Cd, Cu, Ni, Pb, V and Zn) in Budapest

| Pollutant source category | Measurement site | |
|---|---|---|
| | OKI | ELTE |
| *Waste incineration* | 65 | 70 |
| *Traffic* | 17 | 11 |
| *Coal burning* | 5 | 6 |
| *Oil burning* | 6 | 6 |
| *Other sources* | 7 | 7 |

## 5.4.4 A field measurement campaign in Stockholm, Sweden

### *Major scientific issues investigated (Stockholm)*

During three years (1998-2000), data has been obtained in order to evaluate source-receptor relationships of both gaseous and particulate air pollution in the urban area of Stockholm. The project has included both emission and air quality measurement campaigns, and modelling using dispersion models and source receptor models (Johansson *&* 2001).

Detailed emission databases have been created in co-operation with the Environment and Health Protection Administration of Stockholm (URL 5.1). These databases include $NO_x$, CO, PM, benzene and polycyclic aromatic hydrocarbons.

## Relevance to SATURN aims (Stockholm)

The overall scientific objectives were:

- to establish source-receptor relationships for hydrocarbons and PM,
- to evaluate the relative contribution of individual sources for the distribution of these compounds and
- to evaluate emission inventories using a combination of measurements, source-receptor models and dispersion models.

## Experimental set-up (Stockholm)

Three different experimental activities have been carried out:

- a road tunnel study; on-road emissions were characterised,
- a wood combustion experiment; characterisation of the aerosol and some gaseous compounds and
- an urban field campaign in Stockholm.

In the road tunnel study, two sampling points were used: at a distance of 100 m from the entrance and 900 m into the tunnel. Both sites were equipped with Sierra Anderson Hi-Vol $PM_{10}$ inlets, through which the tunnel air was brought through the tunnel wall and into an adjacent space housing the instruments. The following aerosol and gas measurements were performed: aerosol size distributions (from 3 to 800 nm; DMPS), $PM_{10}$ mass (TEOM), coarse and fine fraction elemental composition (PIXE), organic and elemental carbon (ACPM), CO, NO, $NO_2$, light hydrocarbons (C2-C7), volatile hydrocarbons (> C5), alkanes (> C10), PAH's, benzene, toluene, xylene and ethyl-benzene. These measurements (except for PAH) were carried out with a time resolution of between 15 minutes and 1 hour. Furthermore, categorised traffic flows and speeds were recorded as well as the wind velocity, temperature and relative humidity in the tunnel.

For the urban field campaign, the same equipment as in the tunnel was installed at three sites in Stockholm. The sites included a street canyon with around 40 000 vehicles per day, a residential area just outside the centre and a rooftop site in central Stockholm.

## Main results (Stockholm)

In the tunnel, with around 40,000 vehicles per day, hourly average total particle number concentration reach 1 million during morning rush hours. Most of the particles are in the size range from 10 to 60 nm diameter (Kristensson et al. 2000; Johansson & 2001). Total numbers of particles in the size range from 3 to 7 nm tend to decrease significantly during morning rush hours, due to coagulation and deposition on tunnel walls as shown by Gidhagen et al. (2002).

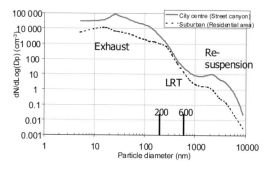

**Fig. 5.8.** Measured particle size distribution in the city centre (upper thick line) and in a residential area (dotted line, lower) in Stockholm. "Exhaust" (particles < 200 nm), "Resuspension" (> 600 nm) and "LRT" (200 – 600 nm) indicate the particle size ranges that are heavily affected by local vehicle exhaust emissions, particles generated by local wear processes and particles which are mainly due to long-range transport, respectively.

The particle size distribution varies considerably from a road tunnel and densely trafficked streets to a background site far from the city. The results suggest that local traffic is the main source of ultra-fine particles and most of the particles close to traffic are in the range from 3 to 30 nm. The size distribution changes markedly as a function of distance from the traffic. At the rooftop location, the size distribution shifts towards larger sizes, compared with the corresponding results at the street level location.

The particle size distributions measured at two different sites in Stockholm were compared with each other. Fig. 5.8 shows three distinct modes in the size distribution: < 200 nm, from 200 to 600 nm, and > 600 nm (as aerodynamic diameters). Local vehicle exhaust particles have a dominant contribution to the number concentration of particles less than 200 nm, whereas long-range transport dominates the sizes between 200 and 600 nm. For the large particles (> 600 nm), local vehicles are again the main source, but these particles are mainly originated from the wear of roads, tyres, brakes etc.

Principal component analysis was performed to aerosol and gas phase data from measurements in a road tunnel. $NO_x$, CO, $CO_2$, VOC's and copper were mainly associated with gasoline exhaust, whereas particulate organic carbon and $NO_2$ shows high loadings on both the gasoline and diesel factor. Particle number concentration is dominated by particles with a diameter around 20 nm and associated with diesel exhaust. Elemental carbon, particle surface area (<900 nm) and volume (<900 nm) show highest loadings on the diesel factor. The road dust factor has high loadings of $PM_{2.5}$ and a number of elements (Si, Fe, Mn etc.).

Birch wood is widely used as fuel in stoves and boilers of Swedish households. An experiment has been undertaken to characterise the combustion aerosol and gaseous emissions of wood combustion in a small stove (Hedberg et al. 2002). In traffic exhaust, the number concentration of particles was largest at 20 nm, while the number distribution from wood burning ranged from 20 to 300 nm.

### 5.4.5 A field measurement campaign in Hatfield, United Kingdom

*Major scientific issues investigated (Hatfield)*

PM in the atmosphere is present in different size ranges, as low as few nanometer to tens of micrometer, with various shapes (McMurry 2000). The environmental impact of these particles, such as, health hazards, acid rain, global albedo and visibility degradation as well as their fate and transport is basically determined by factors including the size of the particles and their chemical composition, in addition to other environmental variables (McMurry 2000). Therefore, to understand the environmental and health impact of ambient particles it is important to study their chemical composition as a function of their size. Concentrations of several metals and ions have been determined for size-differentiated particles. As different sources emit particles in different size ranges, such a study can also assist in the understanding of source-receptor relationships.

*Relevance to SATURN aims (Hatfield)*

Chemical analysis results of size-fractionated samples of ambient particles will help in arriving at source-receptor relationships for air pollutants. Such data will also assist in apportioning the sources of particles and in determining the key particle size ranges contributing to particulate mass metrics such as $PM_{10}$, $PM_{2.5}$ and $PM_1$.

*Experimental set-up (Hatfield)*

The ambient particle samples in 10 size ranges were collected using a MOUDI impactor from a suburban rooftop site, on the campus of University of Hertfordshire, Hatfield. The samples were collected in August and September 2000, and March 2001. The samples from the both campaigns were analysed gravimetrically and for metals (Pb, Zn, Fe, Ni and Cu) using FAAS and ICPES. The samples from first campaign were also analysed for water-soluble ions ($SO_4^{-2}$, $NO^{-3}$, $Cl^-$) using ion chromatography.

*Main results (Hatfield)*

The results of the chemical characterisation from two sampling campaigns are presented in Fig. 5.9. Figure shows the distribution of ambient particle mass in the measured size fractions. The percentage concentrations of the analysed species in $PM_{10}$, $PM_{2.5}$ and $PM_1$ fractions for August - September 2000 campaign are shown in Fig 5.9a, and for March 2001 campaign are shown in Fig. 5.9b. The general size distribution pattern exhibited by the particulate mass is quite similar during both campaigns. The mass concentration during the campaign of March 2001 is slightly lower compared to the August - September campaign, mainly because during the March campaign, the average wind speed was higher and there was a higher level of precipitation.

The $PM_{10}$ mass and all but iron of the analysed species are dominated by the fine fraction. More than two thirds of $PM_{10}$ mass and over 80 % of zinc and lead are in fine fraction (particles of 2.5 micron and less). The contribution of $PM_1$ is dominant except for Fe and Cu. Around 50-60 % of the $PM_{10}$ mass is accounted by particles in the range of $PM_1$. The overall size distribution of particles is similar for the spring and summer campaigns, except for Fe, where the coarse fraction was more significant during the March measurements.

The dominance of the fine fraction in the particle mass and the analysed species point towards the influence of anthropogenic activities on the suspended PM at this site. As expected, the concentration of lead in the campaigns was quite low and comparable (6.6 in summer and 8.0 ng/m$^3$ during spring).

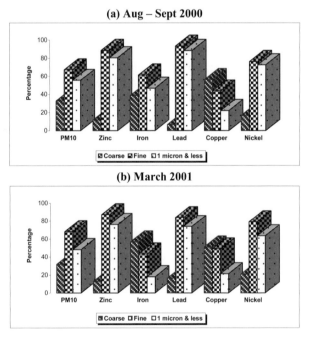

**Fig. 5.9.** Percentage fraction of coarse, fine and $PM_1$ fractions in the metal species and particulate mass during two campaigns (Sokhi & 2001)

# 5.5 Modelling of particulate matter in urban areas

Mathematical models have been developed, evaluated and applied in order to predict aerosol processes, and the number and mass concentrations, such as $PM_{2.5}$ and $PM_{10}$, in urban areas. Source apportionment modelling has been discussed in relation to the field measurement campaigns.

### 5.5.1 Aerosol process models

In Finland, an aerosol dynamical model MONO32 was applied that was originally developed at the University of Helsinki (Pirjola and Kulmala 2000). The model takes into account gas-phase chemistry and aerosol dynamics (nucleation, coagulation, condensation/evaporation and deposition). The particles are classified into four different size modes, which are monodisperse. The first objective was to evaluate quantitatively the influence on aerosol evolution of various chemistry and aerosol processes.

Pohjola et al. (2003) have compiled vehicular exhaust scenarios in selected urban environments. They studied the effects of coagulation, condensation, the concentration of condensable organic vapour, and the dilution of the exhaust plume on the number concentration, composition and particle size.

A particle dynamic module has been developed by Gidhagen et al. (2002). It is intended for studying the dynamics of different particle size distributions in a heavily trafficked urban environment. The model was tested by comparison with measurements in a road traffic tunnel (Kristensson et al. 2001b). A CFD model is coupled to the above mentioned MONO32 aerosol dynamical model. For the tunnel application, a single compound is used to represent the particle composition, although it is possible to handle separately inorganic salts, organic and elemental carbon, sea salt and mineral dust.

The CFD solver is used to calculate the transport equations for the particle number concentration and the particle mass concentration of particle size mode. The effect on particle number concentrations of coagulation and dry deposition was evaluated under morning rush hour conditions. The model simulations showed:

- For the two smallest size modes used in the model (diameter below 29 nm), coagulation had the largest effect on the resulting number concentrations, but dry deposition was also found to be important.
- With the high particle number concentrations that characterise the morning rush hour, coagulation will generate a maximum of the nucleation mode in middle of the tunnel, with decreasing concentration towards the ends of the tunnel.
- For the coagulation process, the hygroscopicity of the particles is not important. A change of sulphate content from 3 % to 50 % in the particles, with resulting

particle growth due to increased water uptake, did not produce any significant changes in the coagulation rate.

- The determination of emission factors of ultrafine particles must, therefore, include the effects of coagulation and wall losses due to dry deposition.
- The transient behaviour of particle number concentrations during 24 hours was also investigated.

From the case study we conclude:

- Constant emission factors cannot explain the observed time series of number concentrations for particles smaller than 29 nm. The model study suggests that the emission factors for light duty vehicles are strong functions of speed; the emissions are reduced when the speed is decreased.
- The total number emission factors of the light duty vehicles (in average 95 % gasoline fuelled cars, the remaining being diesel fuelled) determined from the present tunnel model study ($1.6 \cdot 10^{14}$ to $8.4 \cdot 10^{14}$ veh$^{-1}$ km$^{-1}$) are in general higher than what has been reported from laboratory measurements.
- While translating the estimated particle number emission factors to emission factors for particle mass, the present study yields values comparable to those presented from other tunnel measurements.

If particle number concentration turns out to be important from a health point of view, it will be essential to quantify the emission rates and dynamic behaviour of the most abundant particles, i.e. those with a diameter around 20 nm.

### 5.5.2 Roadside dispersion models

CALINE4 is a US EPA approved roadside dispersion model that incorporates the effects of thermal and mechanical turbulence for roadside situations (Benson 1984; Sokhi et al. 1998). The team at the University of Hertfordshire (UK) has applied the CALINE4 model as well as the ADMS-Urban model to a site at a major motorway (M25). In addition to air quality monitoring, the site also included traffic counts and meteorological measurements.

With regard to $PM_{10}$, CALINE4 and ADMS predictions show good agreement with measured data. In this case the measurements along with the model predictions indicate that the 24 hour running mean standard of 50 $\mu$gm$^{-3}$ will be exceeded. However, this needs further investigation, as there can be considerable contribution from urban background in terms of mass of particles.

Similar work was conducted regarding the model CAR-FMI (e.g., Kukkonen et al. 2001b). This roadside emission and dispersion model was extended to allow for the emissions (discussed in section 5.2.3) and dry deposition of PM. The model predictions were compared with a $PM_{2.5}$ measurement campaign in the vicinity of a major road (Tiitta et al. 2002).

### 5.5.3 Semi-empirical and statistical models

Besides the detailed modelling of aerosol processes, a simple model was developed for predicting the concentrations of $PM_{2.5}$ in urban areas (Tiitta et al. 2002; Karppinen et al. 2002). The influence of primary vehicular emissions is evaluated using the roadside emission and dispersion model CAR-FMI (e.g., Kukkonen et al. 2001b), used combined with the meteorological pre-processing model MPP-FMI.

The regionally and long-range transported (LRT) contribution to the concentrations of fine particulate matter ($PM_{2.5}$) are required for dispersion model computations on an urban scale. However, direct measurements of $PM_{2.5}$ on proper regional background locations are commonly not available. A statistical model for evaluating the regionally and long-range transported concentration of $PM_{2.5}$ has therefore been developed, that can be used for evaluating the fine particulate matter concentrations; based on available measurements at the nearest EMEP (Co-operative programme for monitoring and evaluating of the long-range transmission of air pollutants in Europe) stations (Karppinen et al. 2002).

The modelling system was tested against the results from a measurement campaign in a suburban environment near a major road in Kuopio, Central Finland. The mass concentrations of fine particles ($PM_{2.5}$) were measured simultaneously at four distances from a major road, together with traffic flows and relevant meteorological parameters (Tiitta et al. 2002). This modelling system could also be applied in other European cities for analysing the source contributions to measured fine particulate matter concentrations.

A semi-empirical model has been developed for evaluating the $PM_{10}$ concentrations in urban areas (Kukkonen et al. 2001a). The basic model assumption is that local vehicular traffic is responsible for a substantial fraction of the street-level concentrations of both $PM_{10}$ and $NO_x$, either due to primary emissions or vehicle-driven PM from various surfaces. The modelling system utilises the data from an air quality monitoring network in the Helsinki Metropolitan Area. The model also includes a treatment of the regional background concentrations, and resuspended PM. The model performance was evaluated against the measured $PM_{10}$ data from five air quality stations in the Helsinki area in 1999.

### 5.5.4 Sample results on modelled spatial concentration distributions

#### *The spatial concentration distribution of $PM_{10}$ in Budapest*

The spatial distribution of the $PM_{10}$ concentrations was estimated for Budapest using the ADMS dispersion modelling system (Singles and Carruthers 1999). Example results for one specific day are presented in Fig. 5.10. High-resolution emission inventory included contribution from both point and line sources in the city. Meteorological input of the model was based on data gained from surface meas-

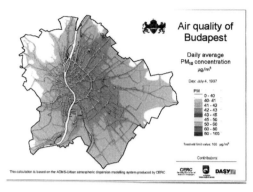

**Fig. 5.10.** Daily average concentrations of $PM_{10}$ in Budapest on 4/7/1997 ($\mu g\ m^{-3}$).

urements as well as on vertical sounding data from Budapest. The model results showed that the highest $PM_{10}$ concentrations occur in the vicinity of major roads in Budapest.

## The spatial concentration distribution of $PM_{10}$ in Helsinki

As an example application for the semi-empirical model developed by Kukkonen et al. (2001a), we have computed the annual average spatial concentration distribution of $PM_{10}$ over central Helsinki in 1998. The predicted distribution is presented in Fig. 5.11.

**Fig. 5.11.** Predicted spatial distribution of yearly means of $PM_{10}$ ($\mu g\ m^{-3}$) in central Helsinki in 1998. The size of the depicted area is 7 km x 7 km. The location of the main streets is shown. The white star indicates the predicted max concentration in the area (41 $\mu g/m^3$)

The computations considered here correspond to urban background concentrations, averaged over the length scale of the grid size applied. The grid interval of the receptor point network in the area considered varies from 50 to 200 m.

Clearly, the concentrations of $PM_{10}$ are highest in the vicinity of the main streets. It is of interest to compare the predicted concentration values with the EU limit values. The limit values in the first and second stages are intended to be in use in 2005 and 2010, respectively; the corresponding numerical values are 40 and 20 $\mu g/m^3$. The first stage limit value would be violated only at the centre of one specific junction of busy streets. However, the second stage limit value is exceeded over a relatively wide area in the vicinity of the main streets of central Helsinki.

### The spatial concentration distribution of $PM_{10}$ in Stockholm

The spatial distribution and population exposure to $PM_{10}$ have been estimated in an earlier project in Stockholm (SHAPE; Stockholm study on Health Effects of Air Pollution and its Economic consequences; Johansson et al. 1999). The calculations are based on an Information and Air Quality Assessment System that is administered by the Regional Association for Air Quality Management in the counties of Stockholm and Uppsala.

Road traffic emissions are calculated from detailed information on traffic volumes, vehicle compositions and driving conditions. All this information is given for short links with a length ranging from less than 100 m up to 2 km. The database also includes emissions from burning of different fuels in district and residential heating, emissions from ships and a special model for airborne re-suspension of PM from road dust. The re-suspension model considers the influence of meteorological conditions, the wetness of the road surface and effect of the traffic on the re-suspension rate.

## 5.6 Exposure to urban particulate matter

### 5.6.1 Variability in exposure

Measurements of road-user exposure to $PM_{2.5}$ in London (Adams et al. 2001a) showed that people moving along city streets are exposed to levels of pollution that are not only substantially higher than urban background fixed-point measurements, but also higher than roadside and kerbside measurements. Volunteers carried high volume portable $PM_{2.5}$ sampling equipment (Adams et al. 2001b) on bicycles, buses, and cars along three routes in Central London, each route approximately 5 km in length. One month of measurements four times a day were made in July, followed by a smaller set of measurement in winter. An additional set of measurements were made by cyclists following their usual commuting routes in the morning and evening.

Day-to-day changes in meteorological conditions, especially wind speed, was the most important determinant of exposure that could be identified using regression modelling (Adams et al. 2001c). Mode of transport was a much weaker determinant, with no detectable effect in the total $PM_{2.5}$ measurements. Isolation of the diesel exhaust fraction of $PM_{2.5}$, using reflectance as a surrogate for elemental carbon aerosol, indicated that car drivers were exposed to higher concentrations than cyclists (Adams et al. 2002).

The most surprising finding was that individual level variability in exposure was as high as the variability due to changes in meteorological conditions. For example, three cyclists following the same route at the same time of the same day measured exposures with geometric standard deviation similar to the geometric mean. This variability is much larger than the possible error in an individual measurement.

### 5.6.2 Modelling road-user exposure and its source apportionment

Dispersion modelling using an operational street canyon plume model in ADMS-Urban was used to investigate likely causes and implications of the observed variability in road-user exposure to $PM_{2.5}$. The street canyon model was run with a high spatial resolution of output receptors along the routes where measurements were made. The modelled exposure was found to be dominated by very high concentrations of $PM_{2.5}$ at the most congested, poorly ventilated canyons along the route, with differences as large as a factor of ten being found between some pairs of road links. This is capable of explaining, in part, the individual level variability in exposure, as small differences in cyclist behaviour including speed lead to differences in the amount of time spent at the most polluted locations. It is also consistent with the observation (Adams et al. 2002) that the individual level exposure is found in the carbonaceous fraction of the $PM_{2.5}$.

Furthermore, the modelling shows how the source apportionment of the $PM_{2.5}$ is very different at the most polluted locations to other parts of the route. Emissions on the nearest road dominate almost to the exclusion of all others at the most polluted points. Along most of the rest of the route, source apportionment is similar to that at roadside locations, with similar contributions from transboundary imported secondary $PM_{2.5}$ to nearest road contribution, and a smaller but significant contribution of primary $PM_{2.5}$ from the rest of London.

Some sources of variability are, however, omitted from this modelling. While the main features of two-dimensional street canyon plume circulation are captured by the semi-empirical canyon model, complex features near junctions are not resolved (e.g. Scaperdas and Colvile 1999). Spatial and temporal variability associated with individual vehicles is also smoothed out, and temporal variability associated with signal control of traffic flows is ignored. It is not known to what extent each of these sources of variability are important in determining variability in exposure. A major programme of work to investigate this in detail in Central London is therefore taking place during 2002-2006 (URL 5.2).

The model results outlined above are all for the mass of PM to which people are exposed. A preliminary study (Bowsher 2000) of the microphysical evolution of aerosol after emission from diesel exhaust indicated that coagulation causes the particle size to increase during the time taken for dispersion to carry the emissions from the vehicle slipstream to the edge of the road. Exposure to ultrafine particles will therefore be even more strongly dominated by short-term peak concentrations than exposure to $PM_{2.5}$.

## 5.7 Conclusions

Several field scale measurement campaigns that focused on aerosols were conducted within the SATURN project. Such campaigns include those performed in Denmark, Finland, France, Greece, Hungary, Sweden and the United Kingdom. A key aspect of these campaigns was the source apportionment of PM. The analysis of these results has contributed to a substantially improved insight on the source contributions, and the chemical and physical characteristics of polluted urban air. For instance, it was demonstrated that local traffic is the dominating source of ultrafine particles, while regional and long-range transport can have a dominating effect on the fine particulate mass ($PM_{2.5}$).

Contributions from various mobile sources may be assessed, e.g., using averaged Scanning Mobility Particle Sizer (SMPS) data in combination with routine monitoring data and traffic rates. Based on the combination of urban sampling, measurement and 3D trajectories, the rural influence on fine particle concentrations and chemical composition can also be evaluated. Chemical mass balance or principal component analysis provide a fairly simple, but useful tool for the assessment of source contributions.

Improved knowledge is needed regarding the chemical and physical properties of urban air particles (especially EC, OC, PAH and metals), surface properties, volatility, hygroscopicity and morphology. It is therefore necessary to apply other advanced analytical techniques, such as SEM and micro probe analysis.

A systematic urban monitoring network is needed in Europe for the evaluation and analysis of fine particles. Such measurements should include not only mass fraction measurements, but also size distribution and chemical composition, in order to achieve reliable information on the origin and potential health effects of PM.

Somewhat smaller resources have been devoted to the mathematical modelling of particulate matter within the SATURN project, compared with those of the experimental work. However, in many cases the measurements were supplemented with one or several modelling methods, such as source apportionment analyses, deterministic and semi-empirical modelling.

Two research teams have performed detailed computations of aerosol transformation, using the aerosol process model MONO32. Interesting results have also been obtained concerning the extension of roadside emission and dispersion models to include a treatment for particle mass fractions. Semi-empirical (partly statistical) models have also been developed for evaluating the concentrations of both $PM_{2.5}$ and $PM_{10}$.

For modelling purposes, a better knowledge is needed regarding the emissions of PM. Information is needed on the number and particle size distribution, and chemical content in vehicular exhaust. For regulatory dispersion modelling, the emission modelling should also be comprehensive in terms of major vehicle categories and driving speeds. Information would also be welcome on the dependence of vehicular emissions, e.g., on cold starts, driving conditions including idling, and fuel composition. Similar information is also needed for biomass combustion, as this is relevant in residential areas in some parts of Europe.

The combined use of aerosol process models and atmospheric dispersion models is another major challenge for the future. Work towards this aim is in progress within the EMEP programme. Evaluating and modelling of the exposure to urban PM is another important topic for future research.

Clearly, atmospheric aerosols have been studied also in many other EUROTRAC-2 projects, such as GENEMIS, AEROSOL, CMD, GLOREAM and PROCLOUD. Closely related work is in progress also within the WMO GURME programme and the COST 715 action.

Urban aerosols will be investigated in several projects of the CLEAR cluster; this is a network of projects that have been granted funding from the "City of Tomorrow" programme of the EU 5th Framework Programme. Several contributors of this review are participants in one or more CLEAR projects; this will therefore be a natural forum in order to utilise and forward the expertise gained in SATURN.

# Chapter 6: Modelling Urban Air Pollution

N. Moussiopoulos[1], H. Schlünzen[2], P. Louka[1]

[1] Aristotle University Thessaloniki, Greece

[2] Meteorological Institute, University of Hamburg, D-20146 Hamburg, Germany

## 6.1 Introduction

Air pollution levels depend on the total emissions, transport and transformation phenomena in the atmosphere and deposition processes. For air quality assessments and for optimising emission reduction strategies all these factors must be considered. The complexity of the overall problem calls for the use of mathematical modelling tools, of so-called air pollution models. Analysing the potential for the practical use of air pollution models implies investigating what kind of statements can be made by the aid of models (qualitative approach) and what is the accuracy of these statements (quantitative approach). Apparently, the former approach is easier, because it does not require more than understanding the characteristics and the range of application of a model. In addition to that, a quantification of the accuracy of model results presupposes insight into (i) input data accuracy and how the latter affects the accuracy of model results, (ii) uncertainties in model assumptions and parameterisations and (iii) methodologies for judging to what extent model results represent reality. As a consequence of the above, model validation by the aid of analytical solutions should be considered as an indispensable part of the model development process, whereas an already validated model should be subject to a genuine evaluation procedure in order to ensure that potential users can assess the degree of reliability and accuracy inherent in the model.

This chapter aims at reviewing the state of the art of local and urban scale air pollution models, with emphasis on those developed and applied in SATURN. Special focus will be on issues related to the multi-scale character of atmospheric processes: Transfer, diffusion and transformation of urban air pollution are governed by processes that occur between the micro/local and meso- scales, while long-term transport mainly concerns the meso- and regional scales. Mathematical modelling deals with all these processes and numerical codes were developed and applied in SATURN to describe phenomena taking place at any of the aforementioned scales, as well as processes resulting from interactions of phenomena between different scales.

## 6.2 Classification of urban air pollution models

Atmospheric phenomena at any specific scale are influenced by the ensemble of interacting atmospheric processes occurring at various scales (Fig. 6.1a). This multi-scale character of the atmosphere is also found for pollutant transport in urban areas (non-shaded scales in Fig. 6.1). The phenomena at the local and urban scales have a horizontal extension of several meters to 500 km and a characteristic time scale of several minutes to several days.

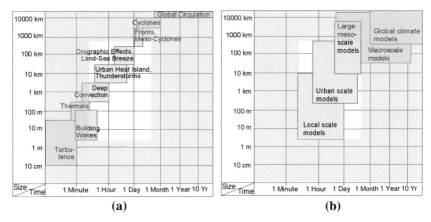

**(a)**                                    **(b)**

**Fig. 6.1.** Characteristic time scale (abscissa) and characteristic horizontal scale (ordinate) of common atmospheric phenomena (a) and of atmospheric models (b) (based on Schlünzen 1996)

Despite the multi-scale character of the atmosphere, practical considerations led to the development of specific-scale models (Fig. 6.1b). In fact, scale separation has been proven to be a quite successful approach for atmospheric modelling, because different approximations and parameterisations can be applied for the different phenomena occurring in the different scales. Specific-scale models should be capable of simulating in detail phenomena occurring in that scale, e.g. an urban scale model should be applicable for simulations of orographic effects, the urban heat island and land-sea-breezes (Fig. 6.1), while the smaller scale phenomena (e.g. turbulence) are parameterised. The type of parameterisation influences the reliability of the model results. On the other hand larger scale phenomena (e.g. cyclones) have to be prescribed in an urban scale model. For this purpose a nesting approach is most successful, which leads to the development of model systems (see section 6.5).

Models can be classified with respect to the scale of the phenomena they are developed to simulate. For the urban area we distinguish the local scale models, which include street canyon models, and the urban scale models. The model domain of a street canyon model is the size of a street canyon. The obstacles (i.e. buildings) are resolved. The model domain of a local scale model is the size of a town or a part of a city (up to 10 km × 10 km). The obstacles (i.e. buildings) may be either resolved ('obstacle resolving local scale models'), or the effect of obstacles may be included implicitly by using specific influence functions ('obstacle accommodating local scale models'). The phenomena that can be resolved by local scale models have to cover a lifetime of at least several minutes and a horizontal scale of several meters. Integration times of one day should be possible to resolve diurnal changes. However, current models only integrate for several hours.

Urban scale models and other larger scale models do not explicitly treat buildings, but include the effect of buildings by using a corresponding roughness length. The model domain of an urban scale model is the size of an EU-country or part of it (up to 500 km × 500 km). The urban scale models may be integrated for several days when using a nesting approach. They can resolve phenomena that have a characteristic time scale more than several 10 minutes and a characteristic horizontal scale of more than several 100's of meters.

The model domain of large mesoscale models covers several European countries (5000 km × 5000 km) and the macro scale models have a model domain that covers part of the globe (larger than 5000 km × 5000 km).

### 6.2.1 Models developed and applied in SATURN

Within SATURN, about 20 models were developed and applied to local and urban scale phenomena. They are listed in Table 6.1 and their features are summarised in a model inventory, which is available on the Internet (Schlünzen 2001, 2002a). The model inventory is based on the information provided by the model contact persons after answering model questionnaires. The received information was first screened and then organised in model overview tables similar to those proposed by Schlünzen (1994). The models were classified with respect to the scale they can be applied to (local scale, urban scale models), to the equations solved (pure dispersion models and combined airflow and dispersion models, henceforth: airflow models) and to the treatment of obstacles (resolving or accommodating).

Airflow models calculate wind, temperature, humidity and concentrations from three-dimensional prognostic equations. This is the case for eight of the selected models (FVM, GRAMM, MEMO, MESO-NH, METRAS, MIMO, MITRAS, SUBMESO, Table 6.2). Models CHENSI and VADIS are not solving the prognostic equation for water vapour. In some cases airflow models include also the diffusion equation for one or several pollutants thus being able to calculate also concentrations. The pure dispersion models prescribe the flow field from other models or use diagnostic equations with empirical factors for treating obstacle influences to calculate the flow field. In one case (OFIS) the shallow water equations are solved for deriving the flow field. Temperature, humidity, etc. are not calculated from prognostic equations, and in most cases a given stratification is imposed. With the exception of OSPM (using a diagnostic equation), all models calculate pollutant concentrations from prognostic equations.

**Table 6.1.** Model names and acronyms, contact persons and selected references

| Model Name | Contact Person | References |
|---|---|---|
| ADMS<br>Atmospheric Dispersion<br>Modelling System – Urban | David Carruthers, CERC Ltd,<br>Cambridge, UK | Carruthers et al. 1997<br>Carruthers et al. 1999a,b |
| CALGRID<br>California Grid Model | Giovanna Finzi, University di<br>Brescia, Dipartimento di Elet-<br>tronica per l'Automazione,<br>Brescia, Italy | Yamartino 1993<br>Yamartino et al. 1992<br>Silibello et al. 1998 |
| CAR-FMI<br>Contaminants in the Air<br>from a Road – Finnish Me-<br>teorological Institute | Jaakko Kukkonen, Finnish Me-<br>teorological Institute, Air<br>Quality Research, Helsinki,<br>Finnland | Härkönen et al. 1995<br>Kukkonen et al. 2001a<br>Öttl et al. 2001 |
| CHENSI | Jean-Francois Sini, Ecole Cen-<br>trale de Nantes, France | Sini et al. 1996 |
| EPISODE | Sam-Erik Walker, Norwegian<br>Institute for Air Research,<br>Kjeller | Gronskei et al. 1992<br>Larssen et al. 1994 |
| FVM | Alain Clappier, LPAS-DGR-<br>EPFL, Lausanne, Switzerland | |
| GRAMM<br>Graz Mesoscale Model | Raimund Almbauer, Dietmar<br>Öttl, Inst. Comb. Eng. &<br>Thermodyn., Techn. Univ.<br>Graz, Austria | Almbauer 1995<br>Almbauer et al. 1995 |
| MARS<br>Model for the Atmospheric<br>Dispersion of Reactive<br>Species | Nicolas Moussiopoulos,<br>LHTEE, Aristotle Univ.<br>Thessaloniki, Greece | Moussiopoulos and Papa-<br>grigoriou 1997<br>Moussiopoulos et al. 1995<br>Moussiopoulos et al. 1997 |
| MECTM<br>Mesoscale Chemistry and<br>Transport model | Heinke Schlünzen, Meteor.<br>Inst., Univ. Hamburg, Ger-<br>many | Lenz et al. 2000<br>Müller et al. 2000 |
| MEMO<br>Mesoscale Model | Nicolas Moussiopoulos,<br>LHTEE, Aristotle Univ. Thes-<br>saloniki, Greece | Moussiopoulos 1989<br>Kunz and Moussiopoulos<br>1995<br>Moussiopoulos et al. 1997 |
| MESO-NH | Robert Rosset, Univ. Paul Sa-<br>batier, Toulouse, France | |
| METRAS<br>Mesoscale Chemistry,<br>Transport and Stream<br>model | Heinke Schlünzen, Meteor.<br>Inst., Univ. Hamburg, Ger-<br>many | Schlünzen 1990<br>Lüpkes and Schlünzen<br>1996<br>von Salzen et al. 1996 |
| MIMO<br>Microscale Model | Nicolas Moussiopoulos,<br>LHTEE, Aristotle Univ.<br>Thessaloniki, Greece | Ehrhard et al. 1999<br>Goetting et al. 1997<br>Kunz et al. 1998 |

**Table 6.1.** (cont.)

| | | |
|---|---|---|
| MITRAS<br>Microscale Chemistry,<br>Transport and Fluid<br>(Stream) model | Heinke Schlünzen, Meteor.<br>Inst., Univ. Hamburg, Germany | Panskus and Schlünzen<br>1997<br>Schlünzen et al. 2002 |
| MUSE<br>Multi-scale for the Atmospheric Dispersion of Reactive Species | Nicolas Moussiopoulos,<br>LHTEE, Aristotle Univ.<br>Thessaloniki, Greece | Sahm and Moussiopoulos<br>1995<br>Sahm and Moussiopoulos<br>1996<br>Sahm et al. 1997 |
| OFIS<br>Ozone Fine Structure<br>Model | Nicolas Moussiopoulos,<br>LHTEE, Aristotle Univ.<br>Thessaloniki, Greece | Moussiopoulos and Sahm<br>1998<br>Sahm and Moussiopoulos<br>1998<br>Sahm and Moussiopoulos<br>1999 |
| OSPM | Ole Hertel, NERI-ARMI,<br>Roskilde, Denmark | Berkowicz et al. 1997 |
| SUBMESO | Patrice Mestayer, Ecole Centrale de Nantes, France<br>Chollet - LEGI, Grenoble<br>Coppalle - CORIA, Rouen | Anquetin et al. 1998<br>Mestayer 1996 |
| UDM-FMI | Ari Karppinen, Finnish Meteorological Institute, Air Quality Research, Helsinki, Finland | Karppinen et al. 2000a<br>Karppinen et al. 2000b<br>Kousa et al. 2001 |
| VADIS<br>Pollutant dispersion in the<br>atmosphere under variable<br>wind conditions | Carlos Borrego, Universidade<br>de Aveiro, Portugal | Martins and Borrego 1998<br>Borrego et al. 2001 |

Pure dispersion models are mainly applied to study transport regimes in the frame of statistical analyses. The models can resolve some spatial differences but most of them give no temporal resolution. They cannot be applied for single case studies. In contrast, airflow models may be applied to study temporal changes in flow and transport regimes for specific situations. The model results are resolved in time and space, but the models need more computer resources. Therefore, they are rarely used for calculating statistical values, however, there is no principal restriction for deriving statistical values from the results of several single case studies.

With respect to their scale of application, the models CHENSI, MIMO, MITRAS, OSPM and VADIS are obstacle resolving street canyon models (Table 6.3). MIMO, MITRAS and VADIS can also be applied to larger local scale phenomena. The local scale models CAR-FMI, UDM-FMI and SUBMESO consider obstacle influences by employing influence functions. The urban scale models ADMS, CALGRID, EPISODE, FVM, GRAMM, MARS, MECTM, MEMO, MESO-NH, METRAS, MUSE, OFIS and SUBMESO treat obstacle influences by using increased roughness lengths.

**Table 6.2.** Prognostic equations solved (•) for momentum (M), pressure (P), temperature (T), humidity (H), liquid water (L), Ice (I), concentrations (C), kinetic energy (E), Dissipation ($\varepsilon$). Calculated variables see table footnote

| Model Name | M | P | T | H | L | I | C | E | $\varepsilon$ |
|---|---|---|---|---|---|---|---|---|---|
| ADMS | | | | | | | Ch | | |
| CALGRID | | | | | | | • | | |
| CAR-FMI | | | | | | | Ch | | |
| CHENSI | u, v, w | | • | | | | • | • | • |
| EPISODE | | | | | | | Ch | | |
| FVM | u, v, w | | $\theta$ | $q_v\rho$ | | | • | • | |
| GRAMM | u, v, w | | $\theta$ | $q_v$ | | | • | • | |
| MARS | | | | | | | • | | |
| MECTM | | | | | | | • | | |
| MEMO | u, v, w | | • | • | | | • | • | • |
| MESO-NH | u, v, w | | $\theta$ | $q_v\rho$ | $q_{lc},\ q_{lr}$ | $q_{si}$ | • | • | |
| METRAS | u, v, w | | $\theta$ | $q_v$ | $q_{lc},\ q_{lr}$ | | • | • | • |
| MIMO | u, v, w | | • | • | | | • | • | • |
| MITRAS | u, v, w | | $\theta$ | $q_v$ | | | • | • | • |
| MUSE | | | | | | | • | | |
| OFIS | $u_{10}, v_{10}$ | | $T_{10}$ | | | | $C_{10}$ | $E_{10}$ | |
| OSPM | | | | | | | | | |
| SUBMESO | u, v, w | • | $\theta$ | $q_v$ | $q_l$ | $q_s$ | • | • | • |
| UDM-FMI | | | | | | | Ch | | |
| VADIS | u, v, w | | • | | | | • | • | • |

*Ch*: hourly concentration values
$E_{10}$: 10 m turbulent kinetic energy
$q_{lc}$: liquid water content in clouds
$q_{ls}$: ice content
$T_{10}$: 10 m temperature

$C_{10}$: 10 m concentration
$q_v$: specific humidity
$q_{lr}$: liquid water content in rain
*u, v, w*: components of wind vector
$\theta$: potential temperature

**Table 6.3.** Classification of local and urban scale models applied SATURN with respect to the equations solved and the treatment of obstacles

| Model Scale | Airflow Model | Pure Dispersion Model |
|---|---|---|
| Local scale | | |
| Street canyon<br>obstacle resolving | CHENSI<br>MIMO<br>MITRAS<br>VADIS (<1 km) | OSPM |
| up to 10km x 10km domain size,<br>obstacle resolving | MIMO<br>MITRAS (<1 km)<br>VADIS (<1 km) | |
| up to 10km x 10km domain size,<br>obstacle accommodating | SUBMESO | CAR-FMI<br>UDM-FMI |
| Urban scale<br>up to 500 km x 500 km domain size | FVM<br>GRAMM (<100 km)<br>MESO-NH<br>MEMO<br>METRAS<br>SUBMESO (<50 km) | ADMS[1] (< 100 km)<br>CALGRID Vers.1.6c<br>EPISODE (<50 km)<br>MARS<br>MECTM<br>MUSE<br>OFIS (<100 km) |
| Large mesoscale, macro scale<br>up to 5000km x 5000km domain size<br>and above | MESO-NH | |

## 6.2.2 Discussion of selected model features

Apart from the above classifications, the most general differences among the models are found in their details, namely, in the use of parameterisations, numerical schemes and nesting procedures. In addition, the treatment of chemistry may differ among the models as well as their requirements in datasets for initial and boundary conditions, which also depend on the nesting used. Last but not least, their status of validation and documentation (Table 6.1) is also quite different.

While vertical exchange is treated in all models (Table 6.4), horizontal exchange is still very much unknown and mostly neglected. In some models (METRAS, SUBMESO) horizontal diffusion is treated via a horizontal filtering approach or the exchange coefficient is set equal to the vertical exchange coefficient (MITRAS). For all other models the model implicit horizontal diffusion is generally unknown and implicitly treated by the horizontal advection scheme.

**Table 6.4.** Parameterisation of the turbulent exchange coefficients

| Model Name | Vertical Exchange Coefficients | Surface Fluxes | Horiz. Diff. |
|---|---|---|---|
| ADMS | traffic included turbulence | Monin-Obukhov | |
| CALGRID | data from CALMET model | | |
| CAR-FMI | van Ulden, Holtslag (85) | | |
| CHENSI | K-$\varepsilon$ | wall functions | |
| EPISODE | measurements, Venkatram | | |
| FVM | K-l | Louis | |
| GRAMM | K-l | Monin-Obukhov | |
| MARS | K-l | | |
| MECTMS | data from METRAS model | | |
| MEMO | K-l, K-$\varepsilon$ | Monin-Obukhov | |
| MESO-NH | K-l, K-$\varepsilon$ | Monin-Obukhov | |
| METRAS | K-l, Countergradient closure | Monin-Obukhov | 7 point |
| MIMO | 0, 1, 2, eq. schemes(K-$\varepsilon$, K-$\tau$, K-$\omega$) | wall functions | |
| MITRAS | K-l, K-$\varepsilon$ | Monin-Obuk.[1] | $K_{hor} = K_{vert}$ |
| MUSE | K-l | | |
| OFIS | algebraic | Monin-Obukhov | |
| OSPM | traffic induced turbulence | | |
| SUBMESO | K-$\varepsilon$, K-$\varepsilon$ -Ri, 5 Eq 2nd order, full 2nd order LES: Smag.-Lilly + K | Monin-Obukhov | Asselin filter |
| UDM-FMI | van Ulden, Holtslag (85) | | |
| VADIS | K-$\varepsilon$ | wall functions | |

[1] wall functions included

Another example for uncertainties in the knowledge on parameterisations is the different treatment of clouds (Table 6.5): the models MESO-NH, METRAS and SUBMESO use prognostic equations for the calculation of liquid water, models EPISODE, GRAMM, MARS, MEMO, MUSE and OFIS prescribe the influence of clouds in a diagnostic way and models ADMS, CALGRID, CAR-FMI, CHENSI, FVM, MECTM, MIMO, MITRAS, OSPM, UDM-FMI and VADIS completely neglect cloud influences or indirectly include them with the prescribed meteorological data.

Concerning pollution transport, Gaussian models (ADMS, CAR-FMI, EPISODE), Lagrangian models (VADIS) and statistical analysis models (UDM-FMI) are applied (Table 6.6). Most models, however, use an Eulerian approach to calculate pollution transport (CALGRID, CHENSI, GRAMM, MARS, MECTM, MEMO, MESO-NH, METRAS, MIMO, MITRAS).

The chemistry modules applied (Table 6.7) range from no chemistry (CHENSI, FVM, MEMO, VADIS) via simple gas phase chemistry considering NO, $NO_2$ and $O_3$ (CAR-FMI, MIMO, OSPM) to complex gas phase reaction schemes (CALGRID, GRAMM, MARS, MECTM, MESO-NH, METRAS, MITRAS,

MUSE, SUBMESO). Aerosols are mostly treated as being passive (ADMS, CALGRID, MUSE, OFIS, UDM-FMI) but some steps are taken to consider chemically active aerosol or aqueous phase chemistry (MESO-NH, METRAS, MITRAS, MUSE). Most models use a resistance model for calculating dry deposition (ADMS, CALGRID, EPISODE, GRAMM, MARS, MECTM, MEMO, MESO-NH, METRAS, MITRAS, MUSE, OFIS), but only very few models consider wet deposition for different pollutants (EPISODE, MESO-NH, METRAS). In respect with the differences in the cloud treatment and aqueous phase chemistry the wet deposition is simulated with quite different complexity in the different models.

The overall model output specifications cannot be generally attributed to the differences in model features. For evaluating the model output a proper model validation is necessary. Most models were compared with measurements or analytic solutions (Table 6.8).

**Table 6.5.** Parameterisation of cloud microphysics and radiative processes

| Model Name | Cloud Parameterisation Scheme | Radiation Budget in the Atmosphere | |
| --- | --- | --- | --- |
| | | Shortwave | Longwave |
| ADMS | | | |
| CALGRID | | estimated | |
| CAR-FMI | | | |
| CHENSI | | | |
| EPISODE | prescribed | | |
| FVM | | mult. reflection in street canyon | |
| GRAMM | diagnostic | • | • |
| MARS | diagnostic | | |
| MECTM | data from METRAS model | | |
| MEMO | diagnostic | emissivity | • |
| MESO-NH | Kessler + ice | Morcrette | Morcrette |
| METRAS | Kessler | 2-stream app. (Bakan 94) | |
| MIMO | | | |
| MITRAS | | | |
| MUSE | diagnostic | | |
| OFIS | diagnostic | emissivity | |
| OSPM | | | |
| SUBMESO | Kessler, Berry-Reinhardt | | Wong |
| UDM-FMI | | | |
| VADIS | | | |

**Table 6.6.** Source types considered, type of transport model and treatment of deposition

| Model Name | Source Types | Transport Model | Deposition | |
|---|---|---|---|---|
| | | | Dry | Wet |
| ADMS | line | Gaussian | Resistance model | washout |
| CALGRID | area, point | Eulerian | Resistance model | |
| CAR-FMI | line | Gaussian | Particles | |
| CHENSI | line, volume | Eulerian | | |
| EPISODE | area, line, point | Gaussian | Resistance model | scavenging |
| FVM | | | | |
| GRAMM | | Eulerian | Resistance model | |
| MARS | area | Eulerian | Resistance model | |
| MECTM | volume, area, point | Eulerian | Resistance model | |
| MEMO | area | Eulerian | Resistance model | |
| MESO-NH | | Eulerian | Resistance model | aq. phase chem. |
| METRAS | volume, area, point | Eulerian | Resistance model | scavenging |
| MIMO | volume | Eulerian | | |
| MITRAS | volume, area, line, point | Eulerian | Resistance model | |
| MUSE | area | Eulerian | Resistance model | |
| OFIS | | | Resistance model | |
| OSPM | | | | |
| SUBMESO | | | | within cloud model |
| UDM-FMI | volume, area, line, point | Statistical analysis | $SO_2$, $NO_X$ | for $SO_2$ |
| VADIS | volume, area, line, point | Lagrangian | Fractional absorbtion layer | |

Prior to SATURN models were individually validated, but their participation in model intercomparison exercises was rather rare. Therefore, SATURN put emphasis on intercomparing models by organising three exercises: one for local scale models and two for the urban scale ones. The intercomparison exercise for the local scale was performed in the frame of TRAPOS (section 6.4), while urban scale models were intercompared in the frame of ESCOMPTE_INT and MESOCOM (section 7.3.3). Only models MEMO and MESO-NH participated in the former exercise, which however started just short before SATURN's completion. On the contrary, all urban scale models (FVM, GRAMM, MESO-NH, MEMO, METRAS, SUBMESO) participated in MESOCOM for two main reasons:

- The test case was relatively simple thus not requiring large resources; it could be treated in parallel to other work without the need of additional funding.
- The set-up of the MESOCOM test case allowed identifying the origin of model output differences; this opened opportunities for fruitful discussions among model developers and users.

**Table 6.7.** Calculation of chemical transformations on-or off-line, photolysis scheme

| Model Name | On-/off-line | Chemical Transformations | Photolysis |
|---|---|---|---|
| ADMS | On-line | NO, $NO_2$, $O_3$, VOC, particles | ● |
| CALGRID | On-line | SAPRC90,CBM-IV, dry passive aerosol | dep. Z-angle |
| CAR-FMI | On-line | NO, $NO_2$, $O_3$, $O_2$ | |
| CHENSI | Passive tracers | | |
| EPISODE | | gas phase | from rad. data |
| FVM | | | |
| GRAMM | Off-line | RADM2, RACM | Madronich |
| MARS | On-line | KOREM, EMEP, RADM2, RACM | dep. Z-angle |
| MECTM | On-line | RADM2 | Madronich, STAR |
| MEMO | | | |
| MESO-NH | On-line | RACM, EMEP, aqueous, phase chem. | Madronich |
| METRAS | Off-/on-line | RADM2, EMEP, sectional aerosol model | Madronich, STAR |
| MIMO | On-line | NO, $NO_2$, $O_3$ | |
| MITRAS | On-line | Reduced RACM | Madronich |
| MUSE | Off-line | KOREM, EMEP, RADM2, RACM passive aerosol | dep. Z-angle |
| OFIS | Off-line | EMEP, aerosol passive | |
| OSPM | On-line | NO, $NO_2$, $O_3$ | ● |
| SUBMESO | Off-line | MOCA (188 reactions, 77 species) | |
| UDM-FMI | | $NO_x$, $NO_2$, inert particles | |
| VADIS | Passive tracers | | |

One outcome of a workshop on the model results was that the selected test case could even have been simpler for an easier identification of the origin of model differences (e.g. models physics, numerical method or overall structure). The more complex test cases are less helpful in this respect and much more resource consuming. However, they allow evaluating how close to reality the model results are.

Overall, model evaluation had a lower priority in SATURN compared to model application or improvement (section 6.3). This can be concluded from the fact that only a few models were subject to either model intercomparison (FVM, GRAMM, MARS, MECTM, MEMO, METRAS, MUSE, OFIS) or the application of an overall evaluation concept (GRAMM, MECTM, METRAS, MITRAS). On the other hand, only a few validation concepts are available for the local and urban scale models. For local scale obstacle resolving CFD models, Panskus (2000) suggested an evaluation concept to check the flow field, which may be also applied to CFD street canyon models. An evaluation concept for urban scale airflow models was suggested by Schlünzen (1997). This concept focuses on the flow and the temperature field as well as on passive tracer transport. Besides these two concepts, validation is only based on single case studies.

**Table 6.8.** Validation and evaluation status

| Model Name | Analytic Solutions | Measurements | Model Inter-comp. | TRAPOS | ESCOMPTE | MESOCOM | Overall Eval. Concept[1] |
|---|---|---|---|---|---|---|---|
| ADMS | | • | | | | | |
| CALGRID | | • | • | | | | |
| CAR-FMI | | • | | | | | |
| CHENSI | • | •[1),2)] | • | • | | | |
| EPISODE | • | • | | | | | |
| FVM | • | | | | | • | |
| GRAMM | • | • | | | | • | • |
| MARS | | • | • | | | | |
| MECTM | | • | • | | | | |
| MEMO | •[3)] | • | • | | • | • | |
| MESO-NH | | | | | • | | |
| METRAS | • | • | • | | | • | • |
| MIMO | | •[1),2)] | • | • | | | |
| MITRAS | • | •[2)] | | | | | • |
| MUSE | | • | • | | | | |
| OFIS | | | • | | | | |
| OSPM | | | | | | | |
| SUBMESO | • | • | | | | • | |
| UDM-FMI | | • | | | | | |
| VADIS | • | •[1),2)] | | | | | |

[1] several different cases
[2] wind tunnel data
[3] individual modules

## 6.3 Urban scale models

Urban scale models are developed to simulate atmospheric phenomena of a horizontal scale below 500 km × 500 km. Six of the thirteen urban scale models applied within SATURN are airflow models (Table 6.3). They simulate the flow, temperature and humidity fields as well as concentrations from prognostic equations (Table 6.2). The other seven models calculate concentrations from a prognostic equation and use flow fields that were calculated with other models.

Within SATURN,

- the existing urban scale models were substantially refined and
- new models and modules were developed and applied (section 6.3.1), as well as validated (section 6.3.2).

## 6.3.1 Model extensions

Several urban scale models were improved and new parameterisations were developed and tested in the frame of SATURN. A primary aim for this development was to ensure the applicability of the models within multi-scale model cascades (see section 6.5).

In this context, work focussed on new methods for considering urban influences on the flow and temperature fields, mainly by improving the parameterisations used to estimate urban heat and momentum fluxes. A good example is the parameterisation implemented in FVM: three active surfaces were considered, namely, the roofs, the walls and the canyon floor. For the momentum, two different roughness lengths were defined for roof and canyon floors, while the contribution of the walls was parameterised with a drag force approach. The sensible heat fluxes were determined as a function of the difference between the air temperature and the surface temperature. A complete energy budget equation was solved for each of the three surfaces. The short and long wave radiative fluxes were computed taking into account the shadows and multiple reflection effects of the street canyon element (Martilli et al. 2000).

Models were also extended to account for microphysical processes and radiative transfer in the presence of single clouds. In MEMO, for instance, three conservation equations for water are being solved for this purpose (water vapour; cloud water and rainwater as fractions of the liquid water content). These three equations are interlinked by the source-sink terms describing the interactions between the three quantities (Fig. 6.2). Suitable parameterisation schemes were adopted for calculating these terms (Kessler 1969). An efficient radiation scheme was embedded in MEMO for providing atmospheric heating/cooling rates and radiative fluxes at ground for both clear and polluted air (Moussiopoulos 1987). The parameterisation of longwave radiation was based on an enhanced emissivity method, whereas shortwave radiation was calculated considering multiple reflections between single clouds. In this way, based on the obtained liquid water content, the liquid water path can be integrated over the height and may be used to calculate the optical depth of the cloud (Stephens 1978). The latter quantity is relevant for calculating the radiative quantities of the cloud.

Other development areas were related to methods providing better urban emission estimates and to improved chemistry modules (e.g. MECTM, Lenz et al.

2000; TAPOM, Calpini & 2001). The latter were first tested in comparison to measured data (Table 6.8) and then applied to investigate, among other, the sensitivity of ozone concentrations in urban areas to the lateral boundary values of concentrations and to emission data uncertainty (Fig. 6.3), as well as the loss of solar radiation by gases and aerosols in the gas phase chemistry calculation.

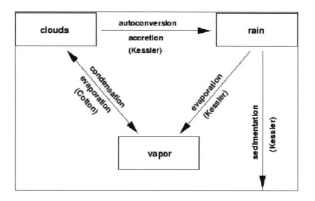

**Fig. 6.2.** The respective interactions between the three water classes

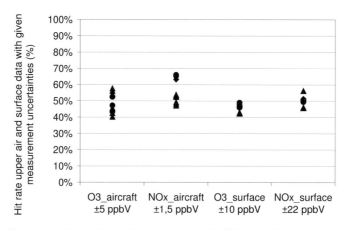

**Fig. 6.3.** Hit rates calculated as relative amount of differences between simulated and measured data, where the difference is below the assumed measurement uncertainty. Differences are given for $O_3$ and $NO_x$ in comparison to aircraft measurements (assumed measurement uncertainty for $O_3$ ±5 ppbv, for $NO_x$ ±1.5 ppbv) and in comparison to surface measure-ments (assumed measurement uncertainty for $O_3$ ±10 ppbv, for $NO_x$ ±22 ppbv). Dots de-note relative amounts for model results using modified lateral boundary values for the $NO_x$ and $O_3$ concentrations, triangles denote relative amounts of differences for model results using emission data changed by ±35%. All model results from model METRAS as applied in Lenz et al. (2000). Figure based on Schlünzen (2002)

Work in SATURN included also the development of new algorithms, as for instance in the case of those related to the treatment of pressure and transport in FVM. The specific algorithm is based on a control volume approach and finite element discretisation techniques aiming at a better grid resolution in the zone of interest (e.g. the city centre) and higher accuracy when the meshes are strongly deformed, for instance in the case of very complex topography (Calpini & 2001).

## 6.3.2 Model evaluation

A focus of SATURN was the validation and evaluation of models (see Chapter 7). This task is very important, since the models used in SATURN are already installed as part of model systems (see section 6.5) that are used to assess, e.g., the health effects of pollutants. Some effort in SATURN aimed at model evaluation, yet – as already stated above – at a lesser intensity compared to the amount of work done on model development and model application. This section summarises experience gained from the evaluation and intercomparison of models, as well as from related sensitivity studies.

The non-hydrostatic mesoscale model GRAMM and the MCCM modelling system were evaluated against measurements obtained during the DATE Graz programme (Dispersion of Atmospheric Trace Elements taking the city of Graz as an example) revealing that model results agree fairly well with the observations for the winter episode, whilst the results obtained for the summer episode showed larger discrepancies compared to measurements. (Almbauer & 2001). The existing model evaluation scheme of Schlünzen (1996) was used to evaluate the CFD models GRAMM (Öttl 2000) and METRAS (Dierer 1997).

As already mentioned in section 6.2.2, the MESOCOM exercise provided the framework for comparing the results of all urban scale models used in SATURN (see section 7.3). Results of the relatively simple MESOCOM test case are given in Fig. 6.4 for concentration values. The discrepancy of the simulated concentrations is very large but still in the range of 20% for the maximum as well as for the minimum concentration values (with the exception of one model). This difference in model results is in the same range as the uncertainty obtained when using one model, simulating the full chemistry and varying lateral boundary and emission data values (Fig. 6.2). Adding up the uncertainties determined from results of different models simulating the same case and the uncertainties determined from the sensitivity studies the resulting overall model uncertainty is about 40%.

As a further important aspect, the effect of thermophysical parameters on the prediction of mesoscale airflow was investigated in SATURN. In particular, sensitivity tests for thermophysical parameters in MEMO were performed by assuming an modifying each parameter separately by ±50%, while at the same time other parameters remained unchanged (Sahm & 2001). The investigation included the impact of surface roughness on the heat balance, variations of the thermal soil conductivity, the evaporation and the volumetric heat capacity parameters and short-wave albedo variations. Table 6.9 summarises the thermophysical parameters that found to obtain optimal performance for MEMO (Sahm & 2001).

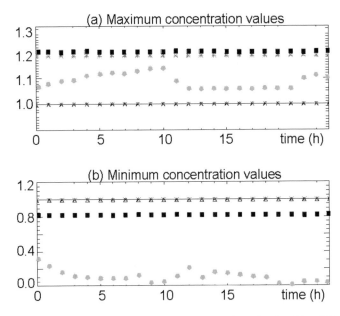

**Fig. 6.4.** Maximum (a) and minimum (b) concentration values simulated by different models for the passive tracer simulation of MESOCOM (Figure based on Galmarini and Thunis 2000)

**Table 6.9.** Land-use types and suggested values of model parameters related to optimal performance of the MEMO model for three different domains.

| Model parameter Land-use type | Roughness length (z0) [m] | Short-wave albedo ( ) | Evaporation parameter ( ) | Volumetric heat capacity (C) [×10⁶ J m⁻³K⁻¹] | Thermal soil conductivity (kB) [×10⁻⁶ m²s⁻¹] |
|---|---|---|---|---|---|
| *Arid land* | 0.01 | 0.11 | 0.01/0.005 | 1.34 | 0.5/1.0 |
| *Few vegetation* | 0.02 | 0.11 | 0.05/0.025 | 1.34 | 0.5/1.0 |
| *Farmland* | 0.02/0.12 | 0.11 | 0.15/0.075 | 1.43 | 0.35/0.7 |
| *Forest* | 0.12/1.00 | 0.05 | 0.20/0.10 | 0.585 | 0.4/0.8 |
| *Suburban* | 0.5 | 0.115 | 0.10/0.05 | 1.10 | 0.65/1.3 |
| *Urban* | 0.8 | 0.10 | 0.05/0.025 | 1.152 | 1.0/2.0 |

# 6.4 Local scale models

Airflow and pollution dispersion within street canyons and city quarters can be described with street canyon/local scale models (section 6.2). Several studies focused in the last years on improving and developing such models. Computational Fluid Dynamics (CFD) codes, especially those applying the standard k-ε model, have been widely applied for predicting local scale airflow patterns. The output of these codes depends, among other, on the numerical scheme used, the way boundary conditions are implemented, the model domain size and the grid resolution. The effects of such parameters on the predicted flow were analysed with detailed investigations of simple cases, such as the flow in a single cavity or around a single mounted cube, as well as more complex configurations. The models' ability in simulating the flow and dispersion characteristics in real cases was also tested. With regard to model quality assurance, both model intercomparisons and encouraging model validation attempts were performed. The latter were based on full-scale measurements and physical modelling in wind tunnels.

A large part of this work was undertaken in the frame of the TMR project TRAPOS (URL 6.1) in which several SATURN groups participated. The result of this research was a significant progress in both the concept and the application range of local scale models. Applications include simulations of the air motion, turbulent field and heat fluxes close to building walls as well as their effect on pollution dispersion. This section is dealing with the main effort undertaken within SATURN related to the synergy in the development of local scale CFD models, operational street air quality models, and their evaluation with the use of atmospheric wind tunnel and street canyon data sets.

The numerical studies within SATURN focused on the issues of (i) the dispersion of passive tracers within the street canyons, with a number of influencing factors such as street geometries, thermal convection due to wall heating, turbulence induced by the motion of the vehicles themselves, and (ii) the emission of pollutants by the vehicles and their rapid transformations as well as development of population exposure models.

## 6.4.1 Microscale influences on dispersion of passive tracers

Over the last years CFD codes have become a standard simulation tool for the analysis, investigation and prediction of the microclimate of urban areas and consequently the pollutant dispersion in these areas. The fundamental problem of CFD simulations lies in the physical difficulties of modelling the effects of turbulence. Other major issues are the accuracy in the spatial discretisation of complex urban geometries, the numerical procedure applied, the boundary conditions and physical properties selected and the validation of the models. These issues were investigated in SATURN with simple or more complex studies.

### Geometrically simple cases

Several studies of airflow in two and three dimensions have investigated simple cases such as single cavities of different dimensions, regular arrays of blocks of buildings and a single cubic building. These studies have demonstrated that local-scale models show reasonable skills in predicting the main features of the flow in simpler cases. In particular, for a single cavity of aspect ratios $W/H$=0.3, 0.5, 0.7, 1.0 and 2.0, the simplest two-dimensional case studied, or a surface mounted cube, the simplest three-dimensional case, the models re-produce sufficiently the pattern of the flow within the cavity and around the cube in accordance with the experimental results, namely main re-circulation and vortices for the cavities, and stagnation point, separation and horse-shoe vortex for the cube (e.g., Sahm et al. 2002; Kovar-Panskus et al. 2002; Sahm $\mathcal{O}$ 2001). This study was run in close co-operation with SATURN and TRAPOS participants.

A CFD model intercomparison exercise performed within TRAPOS elucidated major factors influencing the calculated velocity values, especially at locations where the local flow is strong (Fig. 6.5). Specifically, analysis of flow details, i.e. the local flow adjacent to the walls and ground, and a detailed examination of the source code of each of the CFD models (CFX-TASCflow, CHENSI, CHENSI-2, MIMO, MISKAM) proved the significance of the discrete implementation of the wall-function and of the numerical scheme applied. The deviation between measurements and model predictions was found to be mainly related to the use of the standard k- model. This holds particularly for the calculated turbulent kinetic energy, since the standard k- model assumes that turbulence is isotropic and hence overpredicts the turbulent kinetic energy production within cavities and around the cube (Sahm et al. 2002; Kovar-Panskus et al. 2002).

Small deviations in the calculated velocity and turbulent kinetic energy close to walls and ground may affect pollutant dispersion predictions. Large discrepancies among the model results may also be related to differences in the dispersion modules used as well as to the implementation of the source conditions (cf. Fig. 6.6).

Another CFD study dealt with the airflow in streets with asymmetric walls and different width to height ratios as well as different roof configurations showing the sensitivity of the typical vortex patterns in a canyon to street geometry. This application illustrated the trapping of air in deep narrow streets with transfer of the highest concentrations to the opposite side of the street, and the implication of assuming buildings of equal height on both sides (Assimakopoulos et al. 2002).

The simulation of the airflow between finite arrays of buildings and intersections and the validation of the results with high-quality wind-tunnel data sets are other rather simple cases for which CFD models were applied. For example, the local scale model MIMO (Ehrhard et al. 2000) was validated against available high-quality experimental data obtained in the wind tunnel of the University of Hamburg (URL 6.2). One test case consisted in the simulation of the flow in a finite array of squared shaped rings of buildings. MIMO was found capable of reproducing the observed flow pattern. Specifically, significant departures from the typical vortex structure developing in an infinitely long square street canyon were observed, the main vortex in one of the formed street canyons having its centre shifted upwards and towards the leeward wall. As shown in Fig. 6.7, the vertical

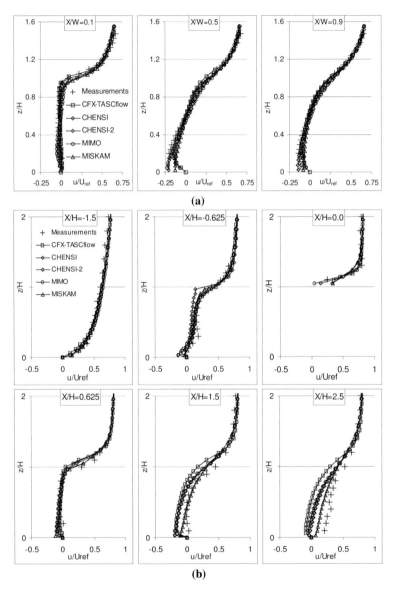

**Fig. 6.5.** Comparisons of the profile of the normalised u-component at different positions **(a)** within a single cavity of aspect ratio W/H=2; X/W=0.5 corresponds to the middle of the cavity, and **(b)** in the centre plane of the flow in the case of the cube; X/H=0.0 corresponds to the position on the top and at the centre of the cube after Sahm et al. (2002)

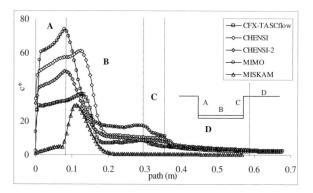

**Fig. 6.6.** Comparison of the normalised concentration c* along the path ABCD for the cavity with W/H=2 (A and C correspond to 75% of the cavity depth, while B and D are equal to its width) after Sahm et al. (2002)

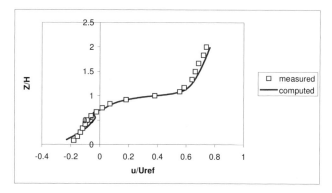

**Fig. 6.7.** Vertical profile of the normalised wind velocity in the centre of an intersection calculated with MIMO compared with corresponding wind tunnel measurements after Assimakopoulos et al. (2002)

profile of the normalised wind velocity obtained with MIMO agrees very satisfactorily with the wind tunnel measurements. The airflow and pollution dispersion in an intersection was investigated using the general purpose engineering code StarCD. The model succeeded in reproducing well the flow direction and speed observed in the wind tunnel, and modelled tracer concentrations in the same street as the source $S$ (Fig. 6.8) were within tens of per cent of the wind tunnel measurement. Good predictions of tracer concentration in the side streets were also made, although errors of a factor of two in the decay rate away from the intersection are sufficient to give concentrations further into the side streets that are wrong by an order of magnitude (Scaperdas et al. 2000, 2001).

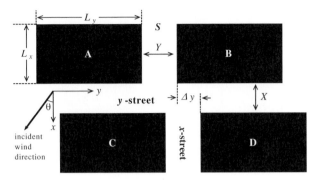

**Fig. 6.8.** Building configuration for the intersection

## Geometrically complex cases

Apart from the simple but useful for extracting information on the abilities of the models cases, microscale prognostic models have accomplished a large progress in predicting complex urban flow fields in real three-dimensional cases (Ketzel et al. 2002; Schlünzen $\mathcal{G}$ 2001). The airflow and pollution dispersion in complex urban configuration with geometrical irregularities were investigated with several models based on real street canyons for which measurements are available. Göttinger Strasse in Hanover, Germany, was one of these cases for which also a

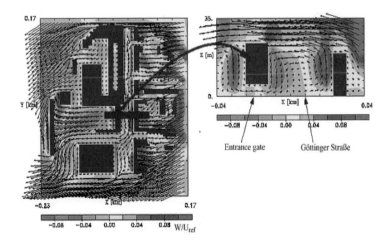

**Fig. 6.9.** Contours of normalised vertical wind speed and vectors of the normalised horizontal wind for a city quarter at 10m height (left) and through the entrance gate (right) (Schlünzen $\mathcal{G}$ 2001)

**Fig. 6.10.** Flow fields for Göttinger Strasse, Hanover calculated with five CFD codes and measured in the wind tunnel for approaching flow from 240° in a horizontal plane 10m above ground (Ketzel et al. 2002)

physical wind-tunnel model was available. The influence of a gateway on the flow field in the street canyon for wind directions nearly perpendicular to the street was investigated with the prognostic numerical model MITRAS. In this case, the model results were plausible; the expected general flow pattern showing a vortex in the street canyon was obtained and compared well with measurements. In the area of the gateway the vortex structure breaks up and the flow continues with a speed up in the gateway, ending up in the vortex in the lee side of the building. (cf. Fig. 6.9). Therefore, such openings as gateways should be considered when simulating pollutant transport at local scale (Schlünzen & 2001).

A CFD model intercomparison performed in the frame of TRAPOS led to valuable results for such complex configurations. In particular, it was shown that the application of CFD codes using the standard k-ε model to real-world cases may lead to large differences in wind patterns (Fig. 6.10) and pollutant concentrations (Fig. 6.11) at locations near building irregularities, such as intersections, gateways, towers and corners. Similar to the simpler cases, the boundary conditions, the numerical schemes and the grid resolution were found to be important parameters for accurately modelling airflow and pollutant dispersion in complex three-dimensional cases, especially close to building irregularities. The accuracy, therefore, of CFD modelling results for locations affected by local gradients should be treated with special care. For practical purposes, an estimation of averages in time (over different inflow situations) or averages in space (to avoid local gradients) may prove more appropriate.

A Lagrangian approach was the basis for the local scale model VADIS for the calculation of instantaneous concentrations of gaseous pollutants. Evaluation with data obtained in Southern France and also against wind tunnel data presented in both cases a good performance and indicated the improvements that the model should be undertaken to better integrate boundary and background concentration values (Borrego & 2001). In Fig 6.12, the comparison between CO measurements from an air quality station and VADIS results is presented (c.f. Fig. 6.12a) as well as the wind and dispersion fields for noon of 16 August (cf. Fig. 6.12b).

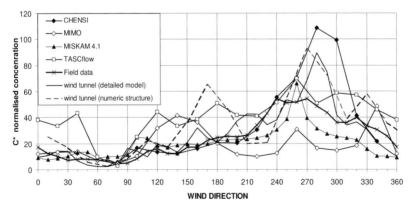

**Fig. 6.11.** Dimensionless concentration calculated with four CFD codes compared with corresponding measurements in the field and in the wind tunnel at a location close to a building corner (Ketzel et al. 2002)

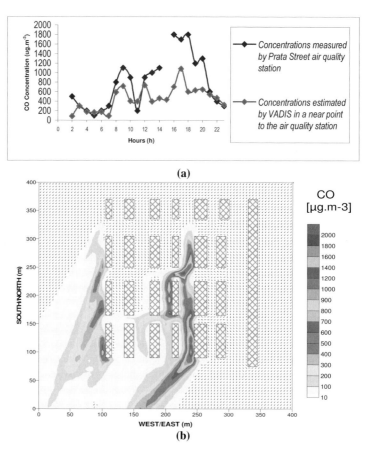

**Fig. 6.12. (a)** Comparison between CO concentrations measured by the station located at Prata Street and calculated by VADIS. **(b)** CO dispersion simulation at noon of 16 August

## *Traffic produced turbulence*

Apart from the geometrical effects, street-canyon flow and vehicles exhaust dispersion are also affected by the vehicles' motion. However, this issue is poorly understood. Only a few full-scale experiments have investigated this topic and have shown that vehicles actually influence at least the vertical transport of air masses in a street canyon. In particular, Qin and Kot (1993) and DePaul and Sheih's (1986) observed a great influence of the movement of the vehicle fleet on the airflow and turbulence within the canyon. Recently, wind-tunnel experiments (Kastner-Klein et al. 2000) showed that the moving vehicles lead to a pronounced transport of pollutants along the canyon axis.

The topic of traffic-produced turbulence has received special attention within SATURN. Modelling research with local scale models focused on introducing modules treating the produced turbulence and the associated parameters affecting it. First approaches consisted in simple parameterisations depending on the vehicle drag coefficient, their number and dimensions as well as the street width (Kastner-Klein et al. 2001) or simple empirical formulae introducing the additional produced turbulence on the basis of field and wind tunnel experiments (Berkowicz et al. 2002; Kastner-Klein et al. 2000). In particular, a dimensionless parameter relating the wind- and vehicle-induced contributions to the turbulent diffusion in a street canyon was proposed for the regimes of diffusion in street canyons. First results of the combined study seem to confirm the validity of the traffic-to-wind turbulence production ratio as a similarity number for a range of turbulent diffusion regimes in a street canyon with moving vehicles, although it can not handle calm situations. The validation of how such influences are implemented in local scale models is the subject of ongoing research. The Nantes '99 experimental results suggest a parameterisation associating turbulence production with traffic density, distance between vehicles and vehicles velocity (Vachon et al. 2002).

Other numerical studies focused on modelling traffic induced flow field and turbulence, traffic emissions and turbulent dispersion in a street canyon and in a tunnel, compared with experimental data obtained in the wind tunnel of the Karlsruhe University. Vehicles were considered as sources of additional momentum and turbulent kinetic energy, which were parameterised through the average vehicle speed and through geometrical parameters describing the traffic (i.e. average vehicle height and average distance between consecutive vehicles). Pollution induced by the traffic was taken into account by considering surface or volume sources of heat, momentum, and mass on the road: the strength of these sources was determined by a global mass, momentum, and heat balance (Hirsch & 1999). An Euler-Lagrangian method based on CFD calculations using Eulerian approach to the continuous phase (air) and Lagrangian approach to the discrete phase of moving objects (vehicles) has been used to investigate traffic induced pollutant production and dispersion. As a basis for the tests, a road tunnel was chosen to assess influence of tunnel length, traffic rate and vehicle speed. This method was applied to road tunnels, street canyons and at intersections (Jicha et al. 2000).

## *Thermal effects*

Apart from street geometry and traffic effects on the formation of the airflow within a street canyon or a road tunnel, buoyant effects due to temperature differences between building facades and air or tunnel jet stream and ambient air, respectively, are also important, especially under low wind conditions. Recently, field campaigns have been investigated such effects enlightening the extent up to which such effects influence the main flow, offering at the same time valuable data sets for the validation of local-scale models (cf. Chapter 4). The numerical approaches are based either on Lagrangian or Eulerian methods.

A simple Lagrangian dispersion model was applied for the description of the buoyancy effects due to temperature differences between jet stream and ambient air, the momentum of the jet stream, and the influence of changing ambient wind directions on the jet stream position. This is the first time that the latter process is considered in a dispersion model for road tunnel portals. Model results were compared with observations of the tracer tests and with results of a microscale Eulerian model using the standard k-ε model. While the new model agreed satisfactorily with the observations, the Eulerian model severely overestimated concentrations in the centre of the jet stream, apparently because it fails to treat the changing position of the jet stream (Almbauer & 2001).

A large part of the research performed under TRAPOS focused on the description of thermal effects on the airflow within a street canyon. Such effects may be associated with the variation of direct solar heating of the street sides and the ground. Implementation of the standard temperature wall function was shown to lead, in this case, to overestimating the influence of wall heating on the airflow pattern. In particular, model simulations failed to reproduce the thin convective boundary layer developing next to the directly heated walls, as for instance observed in the Nantes '99 experimental campaign; instead, the model application led to a largely modified street canyon flow (cf. Fig. 6.13. from Louka et al. 2002). Specifically, if the wall function applied makes use of the temperature gradient normal to the wall, CFD codes cannot resolve thin thermal boundary layers due to grid resolution restrictions. Further investigations are required for solving this problem, an idea being to reformulate the temperature wall conditions e.g. in terms of thermal fluxes based on the thermal balance of the walls.

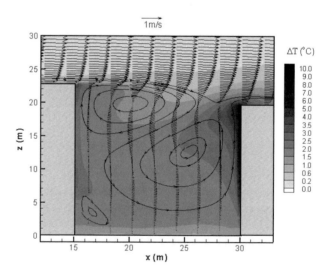

**Fig. 6.13.** Simulated wind field with thermal effects and temperature difference isocontours (after Louka et al. 2002)

## 6.4.2 Chemistry modules and population exposure

Transport and dispersion are not the only processes that determine the relationship between emissions and pollutant concentrations; transformation of pollutants is also important. Thus, chemical reactions with small time scales of the order of a few minutes can have a decisive effect on the composition of the emitted pollutants before scales are reached which are comparable to the resolution of mesoscale models. Photochemical and particle chemistry modules are necessary to be developed and implemented in street-scale models. Such a module is the fast $NO_x$, scale-adapted photo-chemistry including the fast $NO_x$-$NO$-$O_3$ cycle (e.g. Sahm et al. 2000 described in Chapter 8).

$NO_x$ and benzene concentrations have been estimated with inverse dispersion calculations by the operational street air quality model OSPM (Berkowicz et al. 1997). The same model has been used to estimate total traffic emissions and emission factors for the actual structure of the car fleet in St. Petersburg as well as for $PM_{10}$ (Genikhovich & 2001). Benzene concentrations for several streets in Antwerp were computed with the AURORA model. The results were compared with diffusive sampler measurements carried out in 101 street locations during four periods of five days in 1998. Calculated benzene concentrations based on hourly emissions obtained for 1963 road segments in Antwerp showed a very good agreement with the measurements when averaged over periods of 5 days and over all streets (Mensink et al. 2001).

Development of particles modules, modelling of urban particulate matter concentrations and evaluation of models against field-scale data sets as well as development of population exposure models has been the focus of the scientific interest within SATURN. Development and evaluation includes semi-empirical models for $PM_{10}$ (e.g. CAR-FMI, a model for secondary particles; Kukkonen et al. 2001a) and Lagrangian particle models (e.g. GRAL, Graz Lagrangian particle model; Öttl et al. 2001). Intercomparison of local (street) models (GRAM, CAR, CALINE4 and ADMS-urban) for CO, $PM_{10}$ and $NO_2$ with consistent datasets from UK sites suggested that the choice of background values is critical. Two models (ADMS-urban and CALINE4) showed that $PM_{10}$ background was approximately the same as the contributions from local sources (99%ile of 24-hour running means) (Sokhi et al. 1998).

A methodology for assessing urban population exposure on the microscale was developed based on a personal $PM_{2.5}$ sampler and an appropriate exposure model. Results are available for the assessment of road-user exposure to fine particulate air pollution in London using measurements and atmospheric dispersion modelling, specifically a comparison between the two. Related work on dispersion in the urban atmosphere over the short length-scales that are relevant to road-user exposure was also outlined, along with development of capability to model the microphysical evolution of urban aerosols. The road-user exposure with $PM_{2.5}$ measured from the breathing zone of car, bus and bicycle users was compared with ADMS model results. The contributions from nearest road, rest of London and imported were estimated (Colvile & 2001).

The evaluation of the exposure of population with a reasonable accuracy, instead of the personal exposures of specific individuals is important. A model was developed for combining the predicted concentrations, the location of the population and the time spent at home, at workplace and at other places of activity. The computed results are processed and visualised using the Geographical Information System (GIS) MapInfo. The model developed has been designed to be utilised by the municipal authorities in urban planning, e.g., for evaluating impacts of future traffic planning and land use scenarios (Kukkonen et al. 2001b).

Finally, AirGIS was developed for automatic generation of needed input parameters from digital maps and register data for carrying out mapping of street pollution levels e.g. in entire urban areas using the street pollution model OSPM. Various human exposure studies have been and are being carried out in Denmark and serve the basis for the further model development and validation. Recently personal monitoring of particulate matter has been included in these studies and the results indicate that exposure to indoor air pollution plays an important role for the personal exposure to $PM_{2.5}$. Over the next few years the AirGIS system will be applied for calculating human exposures of cohorts in ongoing and historical epidemiological studies in Denmark (Hertel & 2001).

## 6.5 Model systems

Urban air pollution phenomena encompass a wide range of spatial and temporal scales: from a few meters (street canyon pollution) to hundreds of kilometres (secondary pollutant formation in city plumes). Mesoscale wind circulations associated with horizontal temperature gradients, e.g. mountain-valley wind systems and sea/land breezes, particularly affect air quality in cities. In addition, atmospheric circulations created by the city itself, notably the so-called urban heat island, directly influence the dispersion of pollutants. The local concentration levels are also in many cases influenced by regional scale processes such as the atmospheric transport of pollutants emitted in surrounding cities and industrialised areas.

The combined local scale, mesoscale and regional scale effects on urban air quality can be investigated with multi-scale model cascades. The formulation of such cascades has been among the main scientific objectives of EUROTRAC-2 and, in particular, SATURN. Several methods for coupling individual models were proposed, the main aim being to consider the interactions between the various scales and to assess the effect of such interactions on urban air quality.

Specifically, efficient interfaces were developed for linking urban scale model systems to suitable regional scale models. Nested simulations up to the regional scale with a suitable model hierarchy or, alternatively, the use of coherent time dependent boundary conditions allowed the refined description of interactions between larger scale meteorological phenomena and urban scale processes.

One of the main objectives within SATURN was the development of scale exchange procedures in models. Research focussed on the differences in the treatment of various processes at individual scales (microscale vs. mesoscale, urban vs. regional scale) thereby addressing interactions between scale domains. The development of scale exchange procedures, suitable nesting techniques as well as "top-down / bottom-up" model cascade strategies were the main aims. The work, therefore, concentrated on

- formulating methods allowing to deduce from the local scale simulations the parameters needed to represent each city quarter in the simulations at the urban scale, in combination with specific urban mapping outputs, as e.g., terrain pre-processors, urban data base analysers, satellite image processors, or modules preparing boundary conditions.

- assessing the most proper methods of multi-domain simulation of the urban atmosphere with multiple nested grids or variable grids and the development of proper meteorological modules.

- representing adequately local (point or line) sources of reacting pollutants in grid simulations where the grid mesh size is much larger than the source, and to develop sub-grid scale chemical models.

- improving parameterisations of the canopy-atmosphere exchanges and vertical fluxes taking into account microscale effects. As a prerequisite it was thought necessary to undertake studies of the influence of geometrical details, thermal effects, vehicle induced turbulence and fast chemistry.


## 6.5.1 Regional-to-urban coupling

The idea to derive entire model hierarchies by coupling models of different scales is much older than EUROTRAC-2. Yet, only in the last years the scientific community succeeded in considerably refining and further developing nesting techniques, especially for the simultaneous consideration of the regional and urban scales. So, regional scale models have been coupled to global modelling systems, e.g. the EMEP Eulerian photochemistry model.

An example for an integrated multi-scale model is MECTM, nested in results of models of the macro-scale model CTM2 and using meteorological data from the air flow model METRAS (see Fig. 6.14).

Ozone exposure analyses and the assessment of photochemical control strategies call also for regional-to-urban model coupling. In this context, the OFIS model was developed for calculating, in conjunction with a regional scale chemical transport model, urban ozone levels over longer time periods (typically several

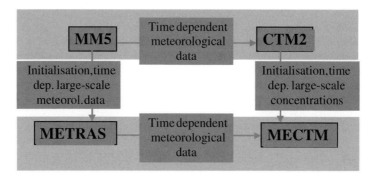

**Fig. 6.14.** Scheme for nesting of meteorological (left) and chemical (right) urban scale models (bottom) into macro-scale models (top)

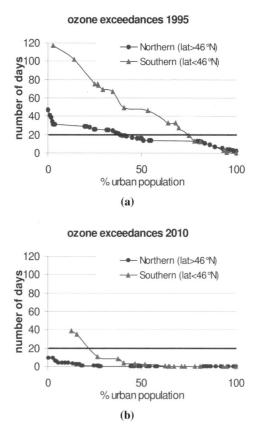

**Fig. 6.15.** Cumulative distribution of days with exceedance of the running 8h-average of 120 µg/m3 ozone over the population in the modelled cities in Northern Europe (circles) and cities in Southern Europe (triangles); **(a)** 1995; **(b)** base case 2010

months). OFIS represents a combination of a box model and a Eulerian multi-layer multi-box model describing transport and photochemical transformation processes in an urban plume (Moussiopoulos and Sahm 2000). Currently, OFIS is extended to account for size and chemically resolved primary and secondary aerosols. Driven with meteorological data and background concentrations calculated with the EMEP MSC-W model for 66 species, OFIS was applied to most large European conurbations in the frame of the "Generalised Exposure Approach", which was EEA's contribution to the EU Commission's Auto-Oil II Programme. As shown in Fig 6.15., this study predicts large air quality improvements in terms of ozone until 2010: Non-attainment of long-term objectives is restricted to only two South European cities (De Leeuw et al. 2002; see also URL 6.3).

## 6.5.2 Urban-to-local coupling

The development of nesting techniques in EUROTRAC-2 is not restricted to the proper one-way coupling of models, i.e. to the use of lower resolution model results for prescribing the boundary conditions of higher resolution models. Especially at the urban-to-local scale range it is also important to deduce quantities needed to parameterise processes at lower resolution from results of higher resolution models. In this context, much progress was achieved in SATURN in the last years, with emphasis on the development of exchange methods between nested models.

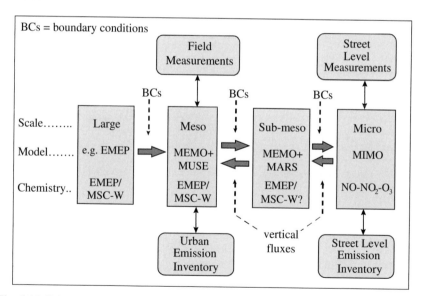

**Fig. 6.16.** Schematic presentation of the multi-scale model cascade in the ZEUS model system (Moussiopoulos ✆ 2001)

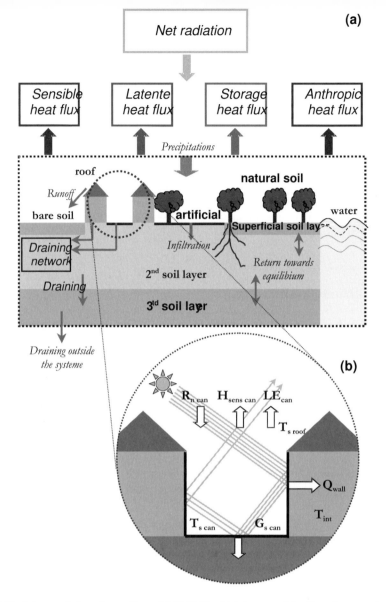

**Fig. 6.17.** Scheme of the urban soil model SM2-U which is part of the French atmospheric model SUBMESO after Dupont et al. (2002)

As an example for a suitable multi-scale model concept for the urban-to-local scale range, Fig. 6.16. shows a schematic representation of the cascade implemented in the ZEUS model system (Moussiopoulos & 2001). Starting at the regional-to-urban scale, nested simulations of MEMO/MARS, the core models of the EZM system (Moussiopoulos 1995) provide meteorological and pollutant concentration data to a series of domains. The microscale domain - resolving the geometrical details of a part of the city - is then allocated within the finest mesoscale domain and initial and boundary conditions are provided to the microscale model MIMO (Ehrhard et al. 1999). Quasi-steady MIMO computations for a group of representative urban geometries provide hourly averaged wind and pollutant concentration fields. These results allow calculating integrated fluxes of momentum, energy and pollutants as input to the mesoscale.

Another example for a multi-scale model cascade developed in SATURN is the French SUBMESO system. The main objective of SUBMESO is simulating the flows, the turbulent fields, the physics and microphysics, and the transport-diffusion-transformations of reactive pollutants within an urban area. This system includes modules for dynamics (Reynolds averaged or LES), microphysics, terrain and soils, chemistry fed by the tropospheric chemistry model MOCA and a grid generator designed for complex terrain. Among the most important features of SUBMESO are a novel meteorological pre-processor, advanced turbulence closure schemes for the stable atmospheric boundary layer (Abart et al. 2000), an external module for the management of nested domain simulations (Pénelon et al. 2002), as well as some new parameterisations for the urban canopy (radiative trapping, heat storage by the walls, mutual shadowing, vegetation over artificial surfaces; Dupont et al. 2000; Fig. 6.17).

An example for subgrid parameterisation in urban areas is the scheme used in the FVM mesoscale model for the estimation of urban heat and momentum fluxes. The parameterisation based on the concept of three active surfaces (cf. section 6.3.1) was implemented in the model and tested against idealised two-dimensional cases. These tests showed that this parameterisation is able to reproduce the main features measured in urban areas for heat fluxes, heat storage, temperatures and momentum fluxes and that the development of the urban boundary layer is very sensitive to these parameters and to their interactions with the city's surroundings (Martilli et al. 2000). This model was also used to simulate the wind circulation over Athens and study the city impact on land and sea breeze (Calpini & 2001).

Low wind situations correspond to critical conditions regarding air quality in several European cities. New, multi-scale modelling approaches are needed for understanding and describing the rather complex phenomenology under such conditions. A good example for new relevant developments within SATURN is the combination of a Eulerian mesoscale model and a Lagrangian particle dispersion model (Öttl et al. 2001). The newly developed Lagrangian dispersion model LDM features random time-steps and a negative intercorrelation parameter for the horizontal wind components. The results of the analyses applying Eulerian autocorrelation functions were compared to measurements made in a suburban area south of the city of Graz. The comparison shows that turbulence under low wind conditions exhibits special effects influenced by the mixed land use and the low building density. Agreeing with sonic anemometer observations, the Eulerian autocorrelation function shows a negative loop when winds are very weak. This

tion function shows a negative loop when winds are very weak. This seems to be an important fact for diffusion under almost calm conditions.

An important application of modelling is also the investigation towards the development of abatement strategies in the context of attaining targets set for the European Air Quality Strategy and Directives. The two most difficult pollutants to consider are $PM_{10}$ and $NO_2$, for both of which it is necessary to superimpose the contributions from local roads or hot spot areas on urban background levels across the city. Thus, the Urban Scale Integrated Assessment Model (USIAM) has been applied to investigate a wide range of abatement scenarios for $PM_{10}$ from transport, based on source apportionment to distinguish contributions from different types of vehicles in different districts or specific roads. There are two areas of London where attainment of standards appears particularly difficult, one in central London, and the other in the vicinity of Heathrow airport to the west of London. A separate study has therefore performed, based on similar principles to the USIAM model, but addressing in detail the limited area surrounding the airport (ApSimon & 2001).

## 6.6 Future perspectives for modelling urban air pollution

The model inventory (section 6.2) documents the status of modelling the urban atmosphere. Improvements of single modules are necessary (e.g. parameterisation of horizontal diffusion, chemical transformations, temperature influences on flow regimes in street canyons, aerosol formation and transport, deposition at building walls) but have to consider the planned application of the model. Not all models need to include all possible model qualities. Therefore, an important parallel step to module improvements has to be the overall evaluation of the models. The detailed evaluation of a model with respect to a defined target value and application range delivers precise information on the necessary model improvements. The application of the different models to several test cases allows distinguishing between shortcomings inherent to all models and shortcomings inherent to single models. Very idealised cases like the MESOCOM test case are essential to determine the reasons of model errors. Based on such overall model evaluations, module improvements and model extensions should be performed and the improved models should be tested further.

While the integration time of local scale models is currently only in the range of hours, future models need to integrate whole days and to consider changes in the meteorological situation, e.g. the partial shading of street canyons in the morning and afternoon. In addition, the time resolution should be increased and large eddy simulations of the local scale need to be performed.

# Chapter 7: Quality Assurance of Air Pollution Models

C. Borrego[1], M. Schatzmann[2], S. Galmarini[3]

[1] Department of Environment and Planning, University of Aveiro, P-3810 Aveiro, Portugal

[2] Meteorological Institute, University of Hamburg, D-20146 Hamburg, Germany

[3] Institute for Environment and Sustainability, Joint Research Center/European Commision, Italy

## 7.1 Introduction

Complex numerical models, widely used to assist decision-makers in environmental studies, have not always been submitted to proper analysis of their performance. Therefore, decisions with political and economical consequences could be based on model predictions of unknown quality (Schatzmann and Leitl 2002). But the definition of model quality is not a simple task and there is no universal approach. Within EUROTRAC-2 it has been recognized that model calculations of *known quality and adequate for their intended use* is an essential element for project success (Borrell 1998).

Evaluation procedures are commonly applied to measure model performance. However, an integrated system of activities should be established in order to guarantee and to improve the quality of modelling results. For this reason, Quality Assurance (QA) plays an increasingly important role among model developers.

The present chapter intends to emphasise the importance of the subject and to present scientific achievements with regard to Model Quality Assurance attained under the SATURN framework. In particular, section 7.2 deals with the fundamentals of Model Quality Assurance, while sections 7.3 and 7.4 address the evaluation of urban and local scale models, respectively.

## 7.2 Fundamentals of Model Quality Assurance

### 7.2.1 Concepts

Model Quality Assurance is a relatively new approach. Nevertheless, Quality Assurance as a general concept is far from being a new term and it is broadly used in many different fields, such as waste disposal and incineration, air and water quality monitoring, laboratory measurements, software development, etc. The main objective of QA is to provide quality assured data, and it requires the outlining of objectives, appropriate tools and an implementation strategy.

A broad definition of QA can be found in International standards on Environmental Management (ISO 1998) and in US EPA documentation (USEPA 1991). A definition of QA is also included in the EUROTRAC-2 glossary (URL 7.1), having the following definitions for air quality modelling:

Model Quality Assurance is an integrated system of management activities involving planning, documentation, implementation and assessment established to ensure that the model in use is of expected quality.

Implementation of QA for modelling introduces a new perspective for estimation of model performance, being judged in terms of usefulness of modelling results or fitness for intended use. To assess quality, Model Evaluation is usually carried out, and can be defined as follows:

Model Evaluation is an overall system of procedures designed to measure the model performance. The following procedures may be considered for Model Evaluation: verification, validation, sensitivity analysis, uncertainty analysis and model intercomparison.

One of the important issues of work within SATURN was to highlight the relation between Model QA and Model Evaluation. In this context and based on the above definitions, Model Evaluation is related to *measuring* model quality, while Quality Assurance is a process to *guarantee* the expected quality. Model Evaluation is therefore one of the inherent components of Model QA (Fig. 7.1).

Model QA is a management function, dealing with policy and designed to ensure that appropriate methods and data are used, errors in calculations are minimised and the documentation is adequate to meet user requirements.

Model Evaluation should have the following components:

- Verification, the process of showing that a technical model is a proper representation of the conceptual model on which is based, and of checking that mathematical equations have been correctly solved. It also includes Quality Control (QC) of the computer code.

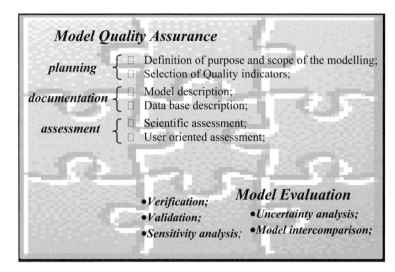

**Fig. 7.1.** Components of Model Quality Assurance

- Validation, the process of showing that the conceptual model and the computer code provide an adequate representation of the problem. It requires comparison of model results with experimental data.

- Sensitivity analysis, the process of identifying the magnitude, direction and form (linear or non-linear) of an individual parameter effect on model results. Sensitivity testing is an important tool that could be used to optimise the QA/QC effort for compiling model input data, primarily for emission data. The identification of pollutants for which model results are highly sensitive will define the need for more accurate estimations for these compounds.

- Uncertainty analysis, the process of characterisation of uncertainties associated to model results. It can be:
  - qualitative (define source and magnitude of error);
  - semi-quantitative (rating, relative indicators);
  - quantitative (statistical parameters, probability method, error propagation).

- Model intercomparison, the process to assess a model performance by simultaneous comparison of modelling results provided by different models for the chosen situation.

## 7.2.2 Quality Indicators and Quality Objectives

Quality of models is understood as a *fitness for purpose* (Britter 1994). Selection of the right model for the defined objectives, together with adequate evaluation procedures, will provide good quality modelling results. Nevertheless, the same model may produce unacceptable results if not used for appropriate conditions. There are no "good" models or "poor" models, the model is suitable or not for the specified objectives.

Due to the complexity of the phenomena simulated by air quality models, there are always uncertainties associated with modelling results. Therefore, an important step is the definition of the criteria to estimate the model performance. A number of statistical parameters are widely used for this purpose, but within these parameters some indicators should be selected as more relevant, depending on the type of answers that the model has to supply. In general, air quality model applications may give rise to the following questions: (1) How well does the model predict maximum pollution levels? (2) How well does the model predict the number of exceedance of the relevant air quality standard? (3) How well do the fluctuations in predictions reproduce the fluctuations of the observed pollutant levels in time and space? (4) How closely are computed concentrations and measurements? For each of these questions, or eventually for other specific situations, indicators of model quality should be chosen (Martinez et al. 1981).

Quality Indicators (QI) reflect the ability of a model to simulate real world phenomena. Mostly, QI are statistical parameters, but in some cases qualitative characteristics, such as representativeness or completeness, can be used. The definition of QI for modelling results is not a simple task. It should be taken into consideration that no single quality indicator is good enough to assess model performance, and that therefore a system of quantitative parameters has to be identified. Application of such indicators helps to understand model limitations and provide a support for model intercomparison.

As mentioned above, QI are selected in accordance with modelling goals. For example, the following statistical parameters are recommended to estimate the UAM model performance (USEPA 1991): unpaired highest prediction accuracy for peak concentration, normalised bias and gross error of all pairs >60 ppb.

It is important to establish whether the uncertainties make the modelling results not useful for answering specific questions. Estimation of model acceptability (Quality Objectives) is based on the definition of the Quality Indicators' range within which the modelling results may be considered satisfactory. For example, for UAM the following ranges should be adopted: normalised bias: ± 5-15 percent; gross error: 30-35 percent.

These values should be realistic and determined primarily according to modelling goals, and should reflect state-of-the-science, or present-day models capabilities. Whenever simulation results do not lie within the recommended range they are considered unacceptable for the specific application. Note the distinction between unacceptable results and unacceptable models.

Analysis of modelling results based on the application of Quality Indicators (relevant statistical parameters) and Quality Objectives (range of acceptability) have been performed for the Lisbon Region (Borrego et al. 1998). Two numerical systems SAIM/UAM (MAR-IV) and MEMO/MARS were applied and evaluated against the same statistical parameters. The results for the ozone concentration are presented in Table 7.1.

**Table 7.1.** Example of QI and QO application to photochemical modelling in the Lisbon Region

| Quality Indicators | SAIMM/UAM | MEMO/MARS | Quality Objectives |
|---|---|---|---|
| Normalised bias | -13 | -2 | ±5-15 % |
| Gross error of all pairs >180 ppb | 27 | 37 | (>60 ppb) 30-35% |
| Unpaired highest prediction accuracy | -2 | 1 | ±15-20 % |

**Table 7.2.** Modelling Quality objectives established by European Directives

| Pollutant | Quality Indicator | Quality Objective | Directive |
|---|---|---|---|
| SO$_2$, NO$_2$, NO$_x$ | Hourly mean | 50-60% | 1999/30/EC |
| | Daily mean | 50% | |
| | Annual mean | 30% | |
| PM, Pb | Annual mean | 50% | |
| CO | 8-hour mean | 50% | 2000/69/EC |
| Benzene | Annual mean | 50% | |

The results show that the models have a good performance in predicting the magnitude of hourly highest concentration (unpaired highest prediction accuracy is about 1-2%). However, the spatial location of the highest concentration is quite difficult to simulate by these numerical systems, as can be seen from the pairs analysis for the ozone concentration higher than 180 ppb. In fact, the gross error for these pairs is 27-37%. It should be noted that the Quality Objectives presented in the table and recommended for UAM were developed for the pairs higher than 60 ppb and not 180 ppb that were used in this analysis for comparison with regulatory limits.

Under the Framework Directive 96/62 for air quality, modelling is considered as a complementary tool to be taken into account for the definition of critical zones and to localise the monitoring points. The Quality Objectives are established in order to determine the acceptability of modelling results. In this context, the uncertainty for modelling and objective estimation is defined as the maximum deviation of the measured and calculated concentration levels, over the period considered, by the limit value, without taking into account the timing of the events. The Quality Objectives for Air Quality Modelling are presented in Table 7.2.

### 7.2.3 The use of models

One of the crucial issues in order to obtain good quality results is the correct application of a model, for the appropriate conditions, and with a correct interpretation of modelling results.

Due to the considerable diversity of available models, it is important to perform their classification. The Model Documentation System (MDS) was built by the European Topic Centre on Air Quality (Moussiopoulos et al. 2000) with the aim of providing guidance to any model user in the selection of the most appropriate model for an intended application. In the MDS information can be found on the technical characteristics, limitation of the models, policy issue and other useful information for potential users.

It is important to realise that models are not always used for the purpose that they were developed. One of the more frequent examples is the situation of models developed to improve the scientific understanding of some phenomena that are used for regulatory purposes. The difference between models developed by researchers for researchers, and those developed for non-expert use were reported by Model Evaluation Group (MEG 1994) and are presented in Fig. 7.2.

The use of models for regulatory purpose implies different requirements in terms of reliability and presentation than those developed for scientific uses only (MEG 1994). The availability of complete documentation for a model is also one of the requirements. However, the current status of model documentation, as assessed by EEA (Moussiopoulos et al. 1996), demonstrates that models used for different policy issues, such as Climate Change, Tropospheric Ozone, Urban Air Quality, etc., have rather good scientific documentation but less complete users manuals. Information on model limitations (definition of the situations for which the results of the model may not be considered as valid) is essential, and should be

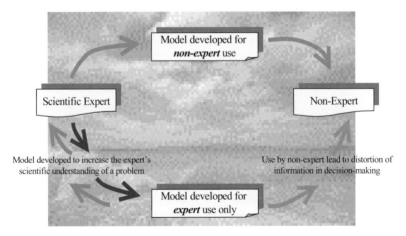

**Fig. 7.2.** Different conceptions of the use of models (from MEG 1994)

available together with the model description. For this purpose a regulatory model should be evaluated under different simulation conditions with regard to its applicability (Schlünzen 1997).

The quality of data used for modelling is another subject for discussion. The total model uncertainty is composed by model physics/chemistry uncertainties, data uncertainties and stochastic uncertainties (Britter 1994). The question rising here is if the data uncertainty should, or should not, be quantified and presented with model quality? It is clear that a suitable model used with inadequate or incorrect data will produce improper results. Therefore, we should look to the modelling objectives. Is the objective to estimate the quality of the model itself or will the modelling results be used as support for political decision? In the first case the data error could be omitted, but in the second case providing the total model uncertainty is essential, and this information about uncertainties associated to modelling results will be so important as modelling results themselves. Within SATURN, a methodology was developed for analysing all contributions to the uncertainty of modelling studies included in air quality assessments (Moussiopoulos et al. 2001).

Another issue refers to the effective communication of model uncertainties to decision makers, so that model outputs can be correctly interpreted. One of the possible ways is to present the modelling results in probabilistic form. Also, qualitative, as well as quantitative information on the uncertainty can be presented (Fox 1984).

## 7.3 Evaluation of urban scale models

### 7.3.1 Introductory remarks

The increasing concern on air quality in urban agglomerations has grown over the last few decades, the research and development in atmospheric modelling of the circulation, emissions, transport/dispersion and chemical transformation of atmospheric constituents and pollutants. A variety of models has been developed and is used for research and regulatory applications. Modelling air quality at the urban scale requires, first of all, an accurate simulation of the atmospheric circulation in the region considered. Prior to the use of an air pollution model to investigate physical-chemical processes or in the frame of a regulatory application, a thorough evaluation of the model capability to reproduce the atmospheric flow has to be conducted.

Within SATURN much progress was achieved regarding the performance analysis of prognostic mesoscale models, i.e. the appropriate meteorological pre-processors for describing urban scale flow configurations. Despite this progress, there is still a need for a generally accepted evaluation protocol for such models, which should adhere to previously suggested principles (cf. Schlünzen 1997). In a more relaxed fashion, prognostic mesoscale models may be assessed in the course of adequate validation and intercomparison activities. Model validation is intended as direct comparison of model results with experimental evidence, whereas model intercomparison involves model vs. model comparisons. This type of model assessment aims at determining the model uncertainty in qualitative and quantitative terms, thus addressing a cardinal issue for the use of models in support of policy making. Model assessment exercises of this type are usually accompanied by rigorous statistical analysis (Galmarini (2001); Girardi et al. (1998); Graziani et al. (1998a); Graziani et al. (1998b); Hanna (1994); Hanna and Yang (2001); Klug et al. (1992); Mosca et al. (1998a); Mosca et al. (1998b); Peilke (2002)).

As further demonstrated in section 7.4, model validation is complicated by the nature of atmospheric flow and the limited representativeness of data collected in the real atmosphere. A successful model evaluation on one case does not guarantee a general applicability of the model to other studies. In this respect, a more comprehensive overview on the model capability to reproduce the atmospheric flow can only be provided by a combined approach involving the comparison of model results for a specific simulation with:

- Observations collected in the real atmosphere or results of laboratory experiments (e.g. wind tunnel data);
- Analytical solutions obtained for simplified ideal cases;
- The results of other models applied to the same case study.

The remainder of this section deals with a model intercomparison (MESOCOM) and a model validation activity (ESCOMPTE_INT) organised within SATURN by the JRC-Institute for Environment and Sustainability. Representing a slight variation of the above combined approach, they provide additional elements and, in this respect, an original contribution to model evaluation procedures.

### 7.3.2 MESOCOM: Model inter-comparison study on ideal flow simulation

MESOCOM introduced an original approach to the evaluation of prognostic mesoscale models. Starting from the assumption that the majority of existing mesoscale flow models have been evaluated against analytical solutions and compared with measurements collected in the real atmosphere, it was considered that a considerable gap exists between the application of the model to the analytical case study (2D simplified flows) and real cases. Namely models have never been applied to cases that are more complex than those for which analytical solutions exist and simpler than real case simulations. The case proposed for simulation within MESOCOM is a well-defined and controlled ideal case that requires the use of a fully 3D mesoscale model but that is not as complex as a real case study. Within MESOCOM a community of 14 European modelling groups (URL 7.2) were asked to simulate the circulation over the domain presented in Fig. 7.3. The models taking part were all non-hydrostatic three-dimensional models based on primitive equations (only one based on vorticity formulation).

The 50×50 km$^2$ domain selected features a sea-land interface. Over land, two Gaussian mountains of different elevation are present. The horizontal grid covering the domain is made of 25 nodes in the x-y direction. In the z-direction 22 vertical layers are prescribed from 0 to 5000 m. The z-coordinate is terrain following. Flow forcing at the lower and upper boundaries, as well as initial and boundary conditions were prescribed for all models. In particular, a time evolving temperature wave was prescribed for the land and constant temperature for the sea surface. Inflow boundary conditions were imposed for temperature and a roughness length was prescribed for the whole domain. The initialisation procedure was based on a single profile. A synoptic wind of 1 m/s from West was prescribed and the vertical profile at sea level was assumed to have a constant 1 m/s value down to 2000 m and then a linear decrease to reach zero at surface. For potential temperature, a 3 K/km stable gradient is assumed everywhere. The boundary conditions for wind and temperature were fixed only for inflow. No boundary conditions were imposed for terms like the horizontal diffusion. Each group was asked to introduce a passive tracer at model start at each level, whose concentration was used to test mass conservation conditions for all models. A 30-hour simulation was requested from each participating group including a 6-hour spin-up time.

The large number of participants, the domain size and the resolution selected in space and time resulted in the collection of a huge amount of results from each group. Therefore, a specific strategy was adopted for the model intercomparison. It consisted in a series of steps starting with the verification that the conditions imposed for the case set-up were fulfilled by all groups. Records for which deviations existed already at this stage were excluded from the further procedure. The first analysis was conducted on the mass conservation condition and the use of the

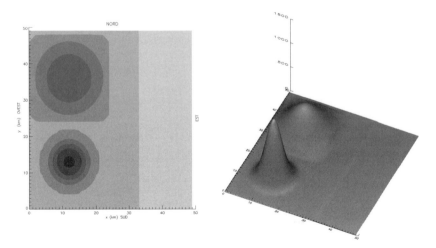

**Fig. 7.3.** MESOCOM domain

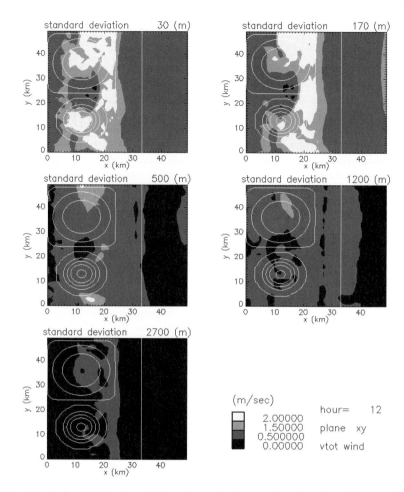

**Fig. 7.4.** Space distribution of the wind velocity standard deviation. Horizontal cress-sections are shown for five vertical levels at 12:00 LST

prescribed surface temperature forcing. Nine models out of the 14 participating to MESOCOM submitted results and all fulfilled the mass conservation. Two models, however, failed to fulfill the requirement connected to the temperature forcing condition. In the next section some results of the intercomparison study for the remaining 7 models are presented. A through analysis of the case study and the full model results can be found in the project website (URL 7.2) and in Thunis et al. (2002).

Examples for results obtained in the framework of the model intercomparison are given in Figs. 7.4 and 7.5. Fig. 7.4 shows the spatial variability of model results (standard deviation) for a fixed time interval (noon) at 5 vertical levels. The variable considered is the horizontal wind component. The figure shows clearly

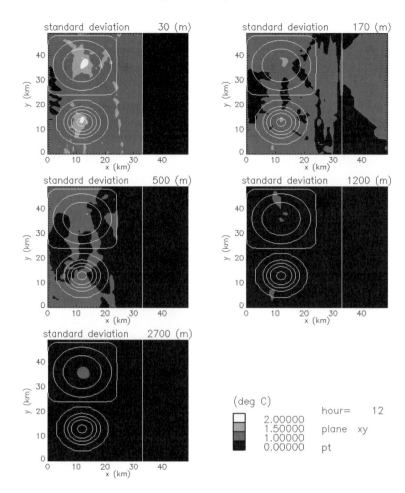

**Fig. 7.5.** Space distribution of the potential temperature standard deviation. Horizontal cress-sections are shown for five vertical levels at 12:00 LST

that at the levels close to the surface the model results differ considerably on the mountain slopes and close to the shoreline where the sea breeze front is located at this time of the simulation. The maximum standard deviation is of the order of 2.5 m/s.

The figure clearly shows that the various models produce a different wind field also in the proximity of the boundaries of the domain over land where the horizontal flow standard deviation is also large. Fig. 7.5 provides the analogous information for the temperature in the same time period. Here again, highest variations arise on the mountain slopes and in the vicinity of the boundaries.

Fig. 7.6 summarises the findings in terms of the standard deviation as a function of time at 5 points selected within the domain for 4 different variables (U wind component, W wind component, total horizontal wind, and potential temperature). Regardless of the location, all variables show a time evolving standard deviation increasing considerably at the on-set of the sea breeze flow. The temperature standard deviation increases monotonically with time.

As an overall conclusion from the MESOCOM activity it was confirmed that model intercomparisons are convenient and useful for gradually and systematically testing model performance.

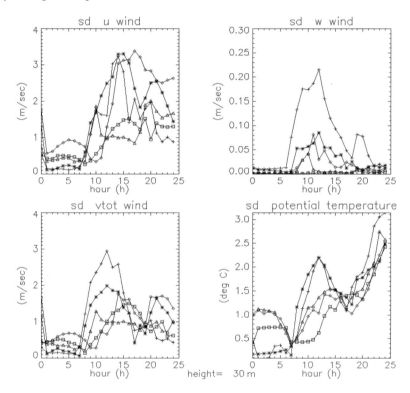

**Fig. 7.6.** Time evolution of the standard deviation of 4 variables at five locations of the domain. Symbols: (plus) top of high mountain, (star) on top of low mountain, (diamond) valley, (triangle) flat terrain between coastline and mountain slopes, (square) shoreline

### 7.3.3 ESCOMPTE_INT: Model assessment on a real case

In the summer 2000 several French and international research institutions coordinated and organised the ESCOMPTE (Expérience sur Site pour COntraindre les

Modèles de Pollution atmosphérique et de Transport d'Emissions) experimental campaign in the region of Marseille (F). The campaign was intended as a preliminary phase of a more extensive and intense campaign to be held in the year 2001. During the so-called 2000 pre-campaign a large number of measurements were collected over a period of one month that relate to meteorology and air quality of the region. The availability of the high quality and detailed database to the international research community stimulated the organisation of the ESCOMPTE_INT model evaluation exercise (URL 7.3). The exercise consisted of the validation of 8 mesoscale models applied to the ESCOMPTE pre-campaign case. The evaluation concerned the flow simulation over the region during a two-day period (30/6-01/07 2000) and the dispersion at the meso-γ scale of four tracers released from 4 different point sources in the region (fictitious releases). Within SATURN it was possible to conduct the validation of 6 models against the collected measurements.

The new aspect of ESCOMPTE_INT is that the simulation case is the only aspect that is common to the participating models. In fact no specification was given concerning the case configuration, boundary and initial conditions and domain size to be selected for the case. Each participant was left free to configure the case in order to obtain the best results with the model used. The reason for that is connected to the intention that any modification to the usual model set up for a case simulation should be avoided, thus allowing the participants to aim at the best simulation possible.

The data for the model evaluation are ground measurements collected at 70 stations distributed in the region (Fig. 7.7). Model-to-model intercomparisons were performed also at 15 additional locations where no measurements were available. The variables evaluated from each model are presented in Table 7.3.

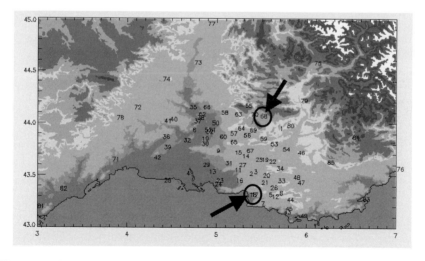

**Fig. 7.7.** Stations of the ESCOMPTE monitoring network

**Table 7.3.** Variables evaluated in each model

|  | U,V,W | PBL h, User defined | PBL h, Richardson number | PBL h, $\Theta$-profile | $\Theta$ | $w \Theta$ (latent and sensible) |
|---|---|---|---|---|---|---|
| At each station as f(t) | X | X | X | X | X |  |
| Vertical profile | X |  |  |  | X |  |
| Spatial distribution |  |  |  |  |  | X |

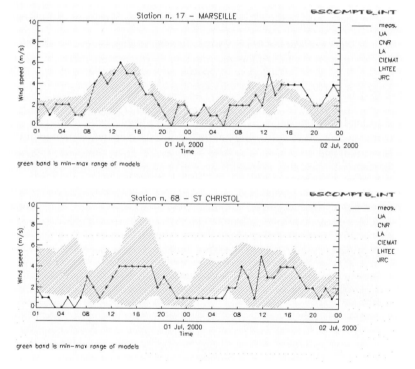

**Fig. 7.8.** Comparison of the time evolution of the total horizontal wind speed at stations 17 (located on the shoreline) and 68 (located in the mountainous zone)

Figs. 7.8 and 7.9 give the result of the comparison of the time evolution of the total horizontal wind speed and direction at two stations (17 and 68) located on the shoreline and in the mountain region of the domain, respectively. The model results are presented globally with no distinctions of single model behaviour. The green band in fact shows the range of variability of the model results over the 48 h period analysed. The line shows the measured values. As far as the wind speed is concerned, models prove to be more capable of reproducing the variable's evolution at the shoreline compared to the mountain region where the range of variability of the results is at some hours as large as 4 times the measured value. The lower panels of Figs. 7.8 and 7.9 reveal a much worse situation concerning wind direction. The figures give for 4 12-hour intervals (represented as concentric circles) the measured wind direction (black segment) and the range of variability of the simulated wind directions (the band has the length of the smallest arc which encompasses all the modelled wind directions). For both stations 17 and 68 the variability in wind direction with respect to the measured one is very large. A behaviour similar to that of the wind speed is found for the temperature time evolution as presented in Fig. 7.10.

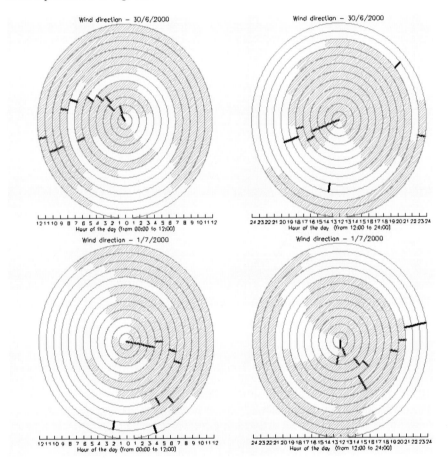

**Fig. 7.9.** Comparison of the time evolution of the wind direction at stations 17 and 68

(Fig. 7.9 cont. overleaf)

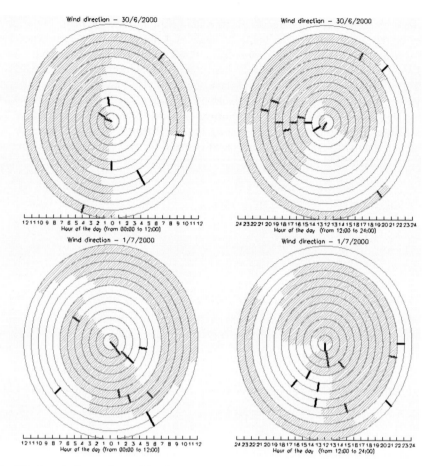

**Fig. 7.9. (cont.)** Comparison of the time evolution of the wind direction at stations 17 and 68

The exercise included also the evaluation of the PBL height determination adopting three different methods. The one used within the model as part of the PBL formulation, the one based on the critical Richardson number and the PBL height obtained from the profile of virtual potential temperature. No measurements are available for direct evaluation purposes. Fig. 7.11 gives for station 17 and 68 the results of the model intercomparison. While the differences station wise are still present, it turns out that the calculations based on the Richardson number give considerably different results in PBL height compared to the other two methods for which an overall agreement is found.

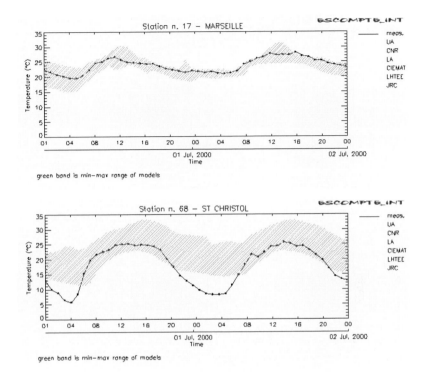

**Fig. 7.10.** Comparison of the time evolution of air temperature at stations 17 and 68

## 7.4 Evaluation of local scale models

### 7.4.1 Introductory remarks

Urban emissions occur mainly within or shortly above the canopy layer i.e. within a zone where the atmospheric flow is heavily disturbed by buildings and other obstacles. It is well known that, in comparison to unobstructed terrain, building effects can change local concentrations by more than one order of magnitude. As a consequence, on the scale of a few streets or city blocks (local scale), it is inappropriate to only consider buildings within a surface roughness parameterisation. Models are needed which are able to resolve the obstacles. Such models are usually hydrodynamic models (CFD codes) which solve the Navier-Stokes equations. Still in use are also the so-called diagnostic models which are based on mass conservation only and describe the effects of the buildings on the flow in a purely empirical manner.

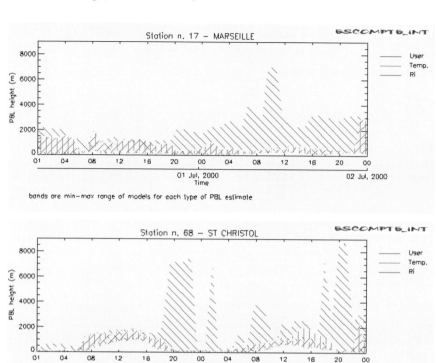

**Fig. 7.11.** Calculated PBL height for the two stations

The subsequent explanations refer to CFD models but similar or identical problems are encountered when dealing with local scale diagnostic models.

### 7.4.2 Problems associated with validation data

Validation comprises a direct comparison between model results and data. As will be shown below, validation of local scale models is not trivial: Data generated in field or laboratory experiments and results from model simulations exhibit systematic differences. To simply compare them with each other is often inappropriate. Special model validation strategies that reflect the particular features of urban canopy layer flows are needed. In the remainder of this section these differences between measurement and calculation will be explained and quantified using the example of data collected at a street canyon monitoring station.

The differences between numerical model results and data are demonstrated using the example of a small area source that continuously discharges a passive tracer into a street canyon (Fig. 7.12, after Schatzmann and Leitl 2002). The figure shows the instantaneous concentrations (in excess above background) as a function of time at the same receptor point and under identical steady-state ambient conditions as they might be found (a) in a field experiment, (b) in a wind-tunnel experiment, or (c) in a numerical simulation with full turbulence parameterisation.

*(a) Field experiments:* High-resolution field measurements usually provide highly intermittent signals, i.e. periods of zero concentration (in excess above ambient) are interspersed with non-zero fluctuating concentrations. It is to be expected that the intermittency of the signal depends largely on the turbulence structure within the canyon and the instantaneous wind direction fluctuations. If the concentration versus time trace varies as shown in Fig. 7.12 (top), long averaging times are required in order to produce a meaningful time-mean-value. It is anticipated that the commonly used 10 min or 30 min measurement cycles are not long enough. Longer averaging times, however, are not usually feasible since the meteorological conditions continuously change during the diurnal cycle. The conclusion is that the repeatability of field results is poor and that large error bars should be attached to time-averaged concentrations determined in field situations as described.

*(b) Laboratory experiments:* When the same dispersion problem is modelled in a wind tunnel, the concentration signal presented in Fig. 7.12 (centre) is obtained. If all the main similarity parameters were matched properly in the small-scale simulation, the time series should resemble that of the field test, but intermittency due to low frequency wind direction variations might be reduced in a ducted flow with insufficient width. Therefore, time-mean concentration maxima determined in laboratory experiments may be larger than those obtained in the field. An important advantage of wind tunnel measurements in comparison to field tests, however, is that the boundary conditions can be carefully controlled, and that numerous repetitions of the same case can be made in order to determine the inherent variability of the dispersing cloud characteristics.

*(c) Numerical model results:* Finally, at the bottom of Fig. 7.12, the concentration versus time trace as obtained from a common grid model is displayed. Provided that the model considers turbulent fluctuations only in parameterised form, in case of constant boundary conditions, it delivers a stationary concentration value. In contrast to the experimental data, this value represents not only a time-mean but also a space-mean concentration representative of the characteristics of the volume of a grid cell.

The example shows that field or laboratory experiments in comparison with the numerical simulations represent distinctively different realities. To compare the results with each other resembles the proverbial comparison of apples with oranges if these fundamental differences are not properly taken into account.

The common approach to solve the problem is to average over the measured concentration time series (c = f (t) in Fig. 7.12, top), thereby applying the same averaging period as is used in the derivation of the assumed (quasi-) steady ambient conditions (e.g., 30 min). However, as will be subsequently shown, concentration mean values based on 30 min averaging intervals are not long enough to obtain representative results. If one compares several half-hourly concentrations measured under nearly identical ambient conditions at the same monitoring station, differences are found which can be as large as an order of magnitude.

This is demonstrated in Fig. 7.13 using the example of data taken over a full year at the street monitoring station Göttinger Strasse in Hanover/Germany by the Lower Saxony State Agency for Ecology (NLÖ 1994, 1995). Shown are normalised concentrations $c* = C\, u_{ref}\, H/(Q/L)$ as a function of wind direction, with C the 30-min averaged measured concentrations at the location of the monitoring station, $u_{ref}$ the wind velocity taken in a measurement height of 100 m, H a characteristic height of the buildings surrounding the street canyon, and (Q/L) the total strength (kg/(m s)) of parallel line sources representing the traffic lanes. Values obtained during low traffic and low wind periods were omitted since under those conditions neither the line source concept nor the assumed neutral stratification are justified.

In theory, all points shown in Fig. 7.13 should fall onto a single curve, but in reality this is not the case. The scatter of data points is caused by the fact that averaging periods of 30 min are simply not long enough to derive representative mean values from strongly fluctuating and intermittent concentration time series. To average over longer time intervals is not however an option since both the meteorological and the traffic conditions continuously change.

The hypothesis that mean concentrations obtainable in urban dispersion experiments are only random samples and not suitable for validation purposes has been verified in corresponding wind tunnel experiments. In a 1:200-scale model of the same site the concentrations at the position of the monitoring station were measured using a Fast Flame Ionisation Detector with a frequency response of approximately 400 Hz. This high resolution in time enabled time series to be collected which were subsequently split into intervals of different length and averaged.

Under the assumption of equal prevailing wind velocities in both the field and wind tunnel, all processes in the wind tunnel are 200 times (=scale factor) faster than in reality. This means averaging intervals of 30 min in the field correspond only to 9 s in the laboratory. Of course, 9 s is much too short to achieve repeatable concentration means. The amount of scatter inherent to short-time averages has been quantified. The difference between the maximum and minimum short time mean value (different depending on wind direction) is in the range of scatter of the data points in Fig. 7.13.

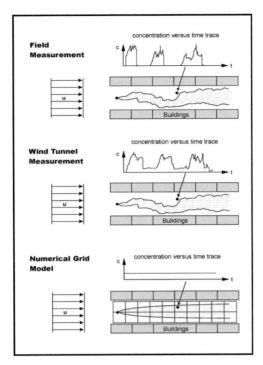

**Fig. 7.12.** Comparison of concentration (in excess above ambient) versus time traces typical for field measurements (top), wind tunnel measurements (centre) and numerical model results (bottom)

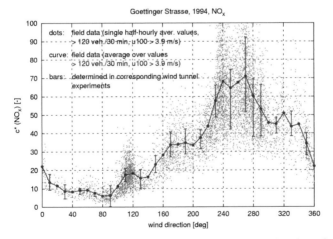

**Fig. 7.13.** Normalized half-hourly mean concentration values as a function of wind direction measured in situ over the period of one year at the street monitoring station Göttinger Strasse in Hanover/Germany (NLÖ 1994, 1995). The error bars were determined in corresponding wind tunnel experiments, see text

The large scatter of data points (small dots) shown in Fig. 7.13 supports the statement that the common 30 min-mean concentrations measured inside the urban canopy layer have the character of random samples only. Depending on the wind direction, the variability between seemingly identical cases can be large. To simply increase the sampling time does not solve the problem, since in the real world the meteorological conditions continuously change. The data points scatter even more than suggested by the wind tunnel results. The reason for this is that in reality the assumption of constant meteorological conditions is already poor for 30 min intervals. To simply increase the sampling time would not solve but worsen the problem since over periods longer than 30 min a systematic trend in meteorological conditions has to be expected.

These findings indicate that for locations within the urban canopy layer which are usually exposed to highly fluctuating and intermittent concentrations, single measurements are not representative. Secondly, these findings demonstrate that the common belief that the uncertainty of field data would be mainly related to the inaccuracy of the instruments may well be false. And thirdly it can be concluded that boundary layer wind tunnel experiments may significantly support the interpretation of field experiments in urban areas. This only marginally increases the total costs of the experiments, and if carried out effectively may fill gaps in the field data and provide substantial help in the analysis and interpretation of this data.

With respect to the validation of local scale obstacle resolving flow and transport models, special strategies need to be applied. In view of the arguments detailed before, it does not make sense to simply compare the results from such models, usually obtained under steady-state conditions, with field data from urban sites, which are random samples from a largely varying ensemble.

To test the quality of obstacle resolving models in a comprehensive way, data sets are needed which are of the same complexity as can be handled by the models. During SATURN such data were made available in the data bank CEDVAL (Leitl 2000, URL 7.4). The speciality of this data bank is that it provides measurements which were taken under steady-state boundary conditions in idealized obstacle arrays of increasing complexity. The following requirements for CEDVAL validation data were postulated:

First, the data set must be complete concerning measured and documented boundary conditions. A validation data set might be called "complete" only if all the boundary conditions of the experiment are measured and documented. These are

a) the geometry of the model or test set-up,
b) the Reynolds number of the flow,
c) the mean wind profile of the approaching flow (representative measurements of all components),
d) the mean profile of turbulence intensity of the approaching flow,

e) the parameters of the logarithmic wind profile (roughness length, displacement height, friction velocity),
f) the spectral characteristics of turbulence of the approaching flow,
g) the integral length scales of turbulence, and for non-neutrally stratified flows
h) a characteristic parameter defining the stratification.

From these input data other parameters as might be needed for complex prognostic simulations can easily be derived. The basic structure of the CEDVAL-data is shown in Fig. 7.14. The data are presently in use world-wide. They provide the means for basic model testing which should be done before the model is exposed to data from real sites.

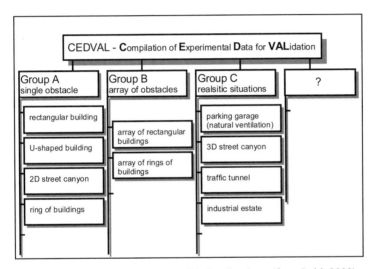

**Fig. 7.14.** Basic structure of the CEDVAL validation data base (from Leitl, 2000)

### 7.4.3 Problems associated with the application of numerical models

Within SATURN, several numerical models were tested against data from a real site, the Göttinger Strasse in Hanover, Germany. One of the models was MISKAM, a prognostic local scale model developed by Eichhorn (1989). It should be emphasised that MISKAM is taken only as an example here, and that the results obtained within the study are representative for other local scale models as well.

MISKAM combines a three-dimensional non-hydrostatic flow field model with an Eulerian dispersion model. It uses a Cartesian grid and the common k-ε turbulence closure scheme and calculates stationary pressure-, velocity- and concentration fields. This model has already been used in several environmental assessment studies. Therefore, it seemed appropriate to test its performance more thoroughly.

Although the model MISKAM uses a non-uniform grid, one has to compromise. While the area covered within the model domain should be as large as possible and the spatial resolution as fine as possible, the model should nevertheless run on a PC within an acceptable duration of time. Schädler et al. (1996) utilized the non-uniform grid capability of MISKAM and used 60 x 60 x 20 grid cells, spread over an area of about 300 m x 300 m x 100m (Fig. 7.15). A view on the site is given in Fig. 7.16, the obstacle composition as it corresponds to the grid and as it was 'seen' by MISKAM is shown in Fig. 7.17.

A comparison of the Figs. 7.16 and 7.17 indicates the degree of geometrical simplification used in the numerical model calculation. The detailed urban structure beyond the immediate environment has been ignored.

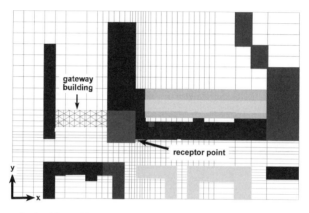

**Fig. 7.15.** Numerical grid used in the MISKAM runs by Schädler (1996). The colours indicate different building heights, see Fig. 7.17

**Fig. 7.16.** View on a section of a wind tunnel model of the Hanover site. The rod indicates the position of the monitoring station

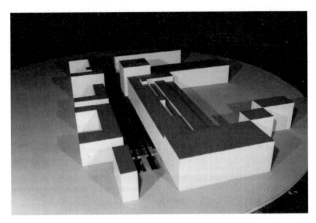

**Fig. 7.17.** Photography of the simplified model structure as it corresponds to the grid in Fig. 7.15

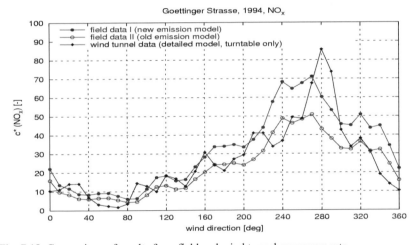

**Fig. 7.18.** Comparison of results from field and wind tunnel measurements

In order to quantify the effect of such geometrical simplifications, a sequence of experiments was carried out. First it was shown that boundary layer wind tunnel experiments are able to replicate the real conditions in good approximation. This is shown in Fig. 7.18 for two emission models as they are presently state of the art. The presentation of results is the same as in Fig. 7.13. In the next step wind tunnel data generated with the detailed physical model were compared with those from the simplified physical model and from the numerical model (Fig. 7.19). As can be seen, geometrical detail really makes a difference. The numerical model prediction is not satisfying. One would have expected that the numerical model matches at least the results from its geometrically equivalent wind tunnel counterpart, but also

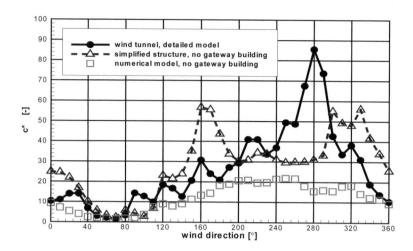

**Fig. 7.19.** Comparison of MISKAM results with those from wind tunnel tests utilizing a scale model with and without geometrical simplification

that is not the case. The maximum concentration from the numerical model is about a factor of 3 smaller than in the wind tunnel experiment.

These results led to improvements in the MISKAM computer code. A number of updates were released, bringing the version number from formerly 1.1 to 3.6. Ketzel et al. (1999) tested the new version of MISKAM against the data, keeping the whole set-up of the run (boundary conditions, choice of grid etc.) the same as in the earlier study. The results are shown in Fig. 7.20. Compared with the results from version 1.1, for some wind directions $c^*$ is now closer to the measurements, for others the changes had an adverse effect. In both calculations, the upper boundary was chosen to be 100 m above ground, approximately 5 times the average building height (H = 20 m) in the area. This choice followed the MISKAM instructions which recommend a domain height of 3 times the building height or more.

In the next numerical tests Ketzel added some grid cells and increased the upper boundary to $z_{max}$ = 500 m. Subsequently the model domain was also enlarged horizontally by adding five empty grid cells around the obstacle array. A general improvement of results is noted (Fig. 7.20) although the shape and the three peaks of the wind tunnel curve are still not reproduced. Ketzel et al. justifiably state that *"this example shows how easily model results can be manipulated by merely varying the choice of parameters which are accessible to the user"*.

The user of complex numerical models should be aware that there are many other parameters with potentially large effects on the model output. For example, it is known that the results of such models are not only sensitive with respect to

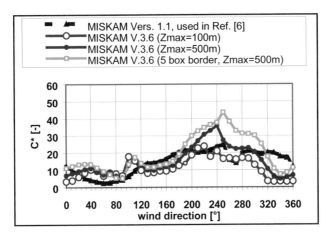

**Fig. 7.20.** Comparison of MISKAM results from different versions of the model and for different set-ups of the numerical grid (from Ketzel et al. 1999)

the grid. Often and even more important is the particular type of internal (at the building surfaces) and external (at the borders of the grid) boundary conditions employed during the model runs.

Another factor that has a large effect on numerical model output is the size of the grid, and not only because it determines the resolution of geometrical details. It is common in urban dispersion problems that the receptor points are close to the sources. In the case of the Göttinger Strasse there were four traffic lanes and the monitoring station was located on a pedestrian walkway directly adjacent to the lanes. The concentration sensor was positioned at a height of 1.5 m. The four different lanes were represented in the numerical model runs by four line sources which emitted into grid cells 2.5 m wide, 3 m high and of variable length. It is a typical feature of numerical grid models that the pollutant flux is uniformly distributed over the source grid cell. This means that for a given emission rate, the initial concentration (the concentration in the source grid cells) is solely dependent on the actual choice of the grid. Doubling the dimension of the source grid cell in the x-, y- and z-directions results in a reduction of the initial concentration by a factor of eight and vice versa. The concentration in grid cells adjacent to the source is to a large degree controlled by the concentration within the source grid cell, i.e. by the actual choice of the grid dimension. It must be concluded that a user who wishes to meet a given data point should, after some grid adjustments, achieve the desired result.

The expectation of obtaining unequivocal results from a given numerical grid model is further undermined by the fact that (for good reasons) most models offer a choice between several boundary conditions, several turbulence closure schemes and sometimes even between different procedures for the numerical solution of the differential equations. This gives the user additional degrees of freedom and makes it unlikely that two users applying the same complex numerical model to the same problem will produce the same results.

This was demonstrated in the European project "Evaluation of Modelling Uncertainty (EMU)" (Hall 1997) in which four experienced user groups predicted the dispersion of dense gas releases around simply shaped buildings by using the same commercially available CFD code. The variability between different modellers' results was shown to be substantial.

A similar conclusion had to be drawn from the TRAPOS model intercomparison study (Ketzel et al. 2002), which was also carried out with the building array of the Göttinger Strasse in Hanover. Although, compared to the study of Schädler et al. (1996), both the spatial resolution and the domain size were significantly increased, the four tested numerical models (all based on nearly the same model physics) predicted quite distinct wind and concentration fields, none of which were in agreement with the measurements.

## 7.5 Summary

It is essential to have a consistent approach to minimise the uncertainty of air pollution modelling and to provide credible results for decision makers. In this outline the Quality Assurance concept has gained increasing attention of both model developers and model users.

Some aspects of Quality Assurance implementation for air quality models have been discussed in the present work. Furthermore, the relation between Model QA and Model Evaluation is highlighted revealing the benefits of the new approach in comparison to the traditional way. To define whether the modelling results are consistently meeting their stated requirements and fit to specified purpose, Quality Indicators and Quality Objectives were established and an example of their application to photochemical modelling was given. Model uncertainty estimation and its communication to the decision makers have been briefly discussed.

The MESOCOM exercise emphasized that in spite of precise conditions imposed on the flow characteristics and set up, the discrepancies among model results are not negligible. Three sources of differences are identified by the exercise. Namely, boundary conditions, surface heat fluxes and horizontal diffusion. In particular the first two were crucial in determining deviation of the flow simulations. The exercise showed that the simple definition of in-flow boundary conditions does not guarantee imposing the same condition for all models since such definition translates into numerical treatments from model to model. Furthermore, although all models used similar formulations to determine sensible heat fluxes from wind and temperature profiles, the former resulted to be very different from the early stages of the simulation. A future exercise of this kind should consider the elimination or the reduction of the influence of these effects, namely by accounting for a larger domain that reduces the impact of boundary conditions and by imposing a time evolution for heat flux at the surface rather than for temperature.

Obtaining reliable predictions of concentration fields inside the urban canopy layer is not straightforward. The numerical models available for this purpose are still relatively new. A model like MISKAM with its frequent updates seems to be still in a development phase. Basic empirical model inputs (turbulence closure scheme, wall functions at internal boundaries) do not appear to be particularly suitable for urban applications. There is a need for further development and, judging from the results of the TRAPOS- or EMU-project, this seems to also be the case for other obstacle-resolving local scale models.

Generally accepted, step-by-step quality assurance procedures for this type of model are urgently needed. A first important step into the right direction has recently been made (VDI 2002). This procedure should come with recommendations for quality assured data sets suitable for validating this type of models. These data sets must be complete in the sense that they encompass (1) both concentration and velocity fields measured at selected positions within larger groups of buildings, and (2) all relevant mean and turbulent velocity parameters defining the boundary layer flow at the edges of the numerical grid. Unfortunately, such data sets do presently exist only for simply shaped single obstacles (CEDVAL). To provide complete data sets also for structured urban sites would be an ambitious task for future research programmes.

Even without the present uncertainties concerning the fitness for purpose of certain model codes and the questions of numerical model accuracy, significant variability remains, arising from the different options the code user has with respect to mesh size, boundary conditions etc. From a regulators point of view this might not be acceptable. Before complex numerical models can become accepted regulatory tools, it is probably necessary to define an additional set of clear instructions that reduces the users' choice to a minimum and ensures a uniform application of the models.

# Chapter 8: Photochemical smog in South European cities

P. Louka[1], G. Finzi[2], M. Volta[2], I. Colbeck[3]

[1] Aristotle University Thessaloniki, GR-54124 Thessaloniki, Greece

[2] University of Brescia, Dipartimento di Elettronica per l'Automazione, Via Branze 38, I-25123 Brescia, Italy

[3] University of Essex, Colchester, UK

## 8.1 Introduction

Photochemical pollution refers to the complex formation of chemical oxidants (mainly ozone-$O_3$) resulting from the interaction of high solar irradiation levels with the precursor substances of nitrogen oxides ($NO_x$), carbon monoxide (CO) and volatile organic compounds (VOCs). $NO_x$ and CO are produced by combustion processes and are emitted by vehicles and industries; VOCs are both anthropogenic and biogenic hydrocarbons (BVOCs). The main BVOCs are isoprene and monoterpenes.

$O_3$ formation and destruction depends upon emissions, concentrations and ratios of precursors (mainly VOC, $NO_x$, and CO), and on the intensity of sunlight. Important in this respect is the role of nitrogen oxide (NO) emissions. Formation of $O_3$ takes place at various space and time scales: the high emission density of reactive precursors in urban areas might lead to high $O_3$ levels within the city or at short distances downwind. At the local and urban scales $O_3$ formation is generally outweighed due to $O_3$ depletion by local $NO_x$ emissions under most meteorological conditions. In urban areas, $O_3$ concentrations may be lower than the rural concentrations due to chemical feeding by local NO (de Leeuw and Bogman 2001). In the suburbs and further downwind of large cities, where local $NO_x$ emissions are lower, the formation generally dominates over depletion and elevated $O_3$ levels are found (de Leeuw and de Paus 2001).

Tropospheric $O_3$ plays an important role also as a greenhouse gas influencing climate change (Kondratyev and Varotsos 2000). It is currently estimated that tropospheric $O_3$ adds 0.4 $W \cdot m^{-2}$ to the current enhanced climate forcing of 2.45 $W \cdot m^{-2}$. The total forcing is a result of the increase in long-lived compounds only ($CO_2$, $CH_4$, $N_2O$, halocarbons) (IPCC 1995). Apart from these climatic effects, tropospheric $O_3$ is a major environmental concern because of its adverse impacts on human health, materials and ecosystems. Human exposure to elevated levels of $O_3$ concentrations can give rise to inflammatory response and decrease in lung function (WHO 1996a). Both laboratory and epidemiological data indicate large variations between individuals in response to episodic $O_3$ exposure. $O_3$ exposure of ecosystems and agricultural crops results in visible foliar injury and in reductions in crop yield and seed production (WHO 1996b). It is also known that $O_3$ affects materials (Lee et al. 1996). Therefore, the issue of $O_3$ formation, its transport and its effects are vital as well as complex, making essential the investigation of the physical and chemical processes that are associated with it in order to develop an international coherent policy to deal with the problem.

In South European countries, with prolonged hot and sunny weather spells during summer, $O_3$ can quickly be formed and high concentrations can occur on many days and in the vicinity of urban centres causing the appearance of severe pollution episodes, while in North Europe the build-up of $O_3$ is slower due to the more moderate weather conditions The relative importance of $O_3$ and its precursors transported into an area characterises a particular episode, related to transport phenomenology, as $O_3$ is highly associated with the presence of $NO_x$ and VOCs. It is therefore necessary to understand how $O_3$ depends on these substances in order to develop an effective policy response (Sillman 1999). In southern European countries exceedances are observed already in April and early May. Later, in June, July and August exceedances are observed all over Europe except in the most northern parts. The lifetime of $O_3$ and/or its pre-cursors in the free troposphere is sufficiently long that pollution may be transported over distances of hundreds to thousands kilometres, resulting in elevated $O_3$ concentrations far from the sources (de Leeuw and de Paus 2001).

Within SATURN the action of FOSEC (Formation of Ozone in South European Cities) was established aiming at contributing towards a better understanding of photochemical pollution in South Europe. FOSEC was organised in a way that ensures maximum possible interaction among various groups within SATURN as well as within other European projects. For this reason, a main element of the organisation of FOSEC was the use of an Internet application, a virtual forum (URL 8.1) that allows the development of its contents and knowledge by people who interact with it. The virtual forum for this action enables the use of electronic tools for exchanging information and summarising the current understanding on urban photochemistry in southern Europe.

In line with FOSEC, this chapter summarises the current understanding on urban photochemistry in South Europe as reflected in the scientific results of various SATURN groups as well as in findings of other European projects. Moreover, some questions for further investigation of this topic are raised. The different sections of this chapter are inter-linked.

## 8.2 Meteorological conditions favouring photochemical smog

It is known that the meteorological influence on photochemical pollution in southern European cities is dominant during every season of the year but especially during summer, and is associated with high sunlight intensity, low winds or stagnant high-pressure systems, high air temperature, high stability (low mixing heights), low midday relative humidity, and occasionally thermally driven mesoscale circulations, primarily sea-land breeze. However, not all high-pressure systems lead to photochemical episodes. Moreover, even in wintertime weather situations may be of relevance for building-up high photochemical pollution levels in southern Europe. Therefore, it is necessary to answer questions like:

1. For which southern European cities is there a synoptic classification which was developed for indicating the occurrence probability of photochemical pollution?
2. Which of these synoptic conditions would lead to the most severe episodes?

### 8.2.1 Synoptic classification

In the summer, the atmospheric circulation in southern Europe is affected mainly by the combination of two global-scale synoptic systems: the subtropical anticyclone of Azores/Bermudas and the west Asia thermal low. The combination of these two synoptic systems provokes the Etesian winds, a semi-persistent wind circulation, consisting of a northerly flow blowing mainly over eastern Mediterranean and especially over the Aegean. The intensity of these winds is controlled by the intensity and the position of the two synoptic systems. When the Etesian winds are strong, the pollution levels are expected to be lower because of the associated strong vertical mixing and diffusion. On the contrary, when the Etesian winds are weak, local circulations prevail due to differential heating. These circulations do not favour the diffusion of the pollutants. This is valid for example in the Athens area, when the land-sea breeze circulation prevails, 'trapping' the pollutants (Kallos et al. 1993; Kassomenos et al. 1998a, b). In addition, due to the presence of the two synoptic systems, large mesoscale circulations develop such as the Iberian, Italian and Anatolian Thermal Lows that also influence the evolution of the regional flows during the day, while pressure differences of up to 30–40 hPa can develop between the Atlantic coast of Portugal and the Arabian Peninsula that favours a general eastward drift of the air masses over the Mediterranean basin (Gangoiti et al. 2001).

In the winter, the depression activity causes a variety of circulation patterns over southern Europe. Some of these patterns are responsible for the accumulation of air pollutants, while some others do not favour high concentrations. The pollution-related synoptic conditions differ for different areas and depend on the topography of each city or region. However, the synoptic conditions generally accompanied by vertical atmospheric stability lead to high pollution levels because of the corresponding temperature inversions and the poor vertical mixing. For example the advection of warm air from the South usually leads to atmospheric stability and is responsible for many pollution episodes during winter. This is also the case when an anticyclone prevails over the region, favouring calm and stable conditions. On the other hand, a strong northerly flow has the opposite results, leading to low pollution levels. The question that has to be answered is whether the above synoptic conditions could be accompanied with sunshine in order to favour also photochemical pollution. This is an object of research and depends mainly on the topography, land-sea distribution and the type of ground cover of each specific region (Kassomenos et al. 1998a).

The Mediterranean is usually divided into three basins, namely, the Western Basin extending from the Iberian Peninsula and from southern France to the northern African coast, the Central Basin from the Adriatic and Ionian Seas to the east coast of Tunisia, and the Eastern Basin including the Aegean and Black Seas (Millán et al. 2002). The synoptic conditions in the Eastern Mediterranean basin and in particular over Greece have been classified in different categories according to the general circulation patterns at the isobaric levels of 850 and 700 hPa. Fig. 8.1 illustrates 8 synoptic patterns over that area. From the pressure patterns

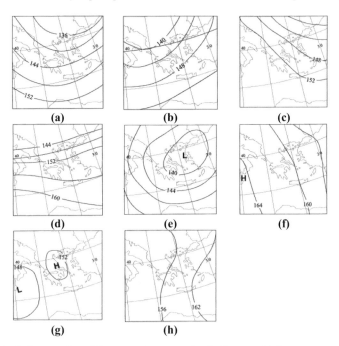

**Fig. 8.1.** Typical examples of the synoptic categories in Greece: **(a)** Long-wave trough **(b)** South-westerly flow **(c)** North-westerly flow **(d)** Zonal flow **(e)** Closed low **(f)** Open anticyclone **(g)** Closed anticyclone **(h)** High-Low (from Kassomenos et al. 1998a)

observed, the anticyclonic circulation has maximum occurrence in January and June, the combination of high-low pressure systems predominates in July and August, situations of cyclonic type dominate in February, March and December, while situations of South-westerly flow dominates in November and April. Weak flows favouring trapping of pollutants are mainly associated with open and closed anticyclones and zonal flow during which mesoscale circulations become dominant (Kassomenos et al. 1998b).

Within the Central Mediterranean basin, synoptic weather classifications derived from the scheme originally proposed by Borghi and Giugliacci (1980), are currently used in Italy for both meteorological and air quality scopes. The classification by Borghi is based on 12 weather types, identified by their geopotential patterns at 850 hPa for the 5-year period 1984-1989 (Fig. 8.2) (Finardi et al. 1999). Following the cited classification the weather types that can most likely favour photochemical episodes in Italy are classes 1, 5, 7 and 8 (c.f. Fig. 8.2). Classes 1 and 8 can favour pollution episodes in parts of the country depending on the position and extent of the anticyclonic circulation.

**Fig. 8.2.** Weather type classification used in Italy (ENEL/Prod ULP 1998)

In the summer, the Western Mediterranean Basin (including Portugal) is usually characterised by a ridge of the Azores High extending over northern Iberia
and southern France. During summer the Azores High expands over the Atlantic
or towards France. In association with the horizontal mesoscale air motion in the
basin, during day-time the pollutants concentrations tend to increase at the northern part of the Iberian eastern coast, while the night-time flow transports the pollutants towards the South (Gangoiti et al. 2001). Most of the Portuguese $O_3$ episodes appear under the presence of the Azores anticyclone extended in ridge over
the Iberian Peninsula, promoting a continental dry and very hot circulation over
Portugal. These synoptic conditions are characterised by slightly above average
mean sea level pressure, almost non-existent surface pressure gradients, and is
generally associated with weak winds in the lower troposphere, cloudless skies,
high maximum temperatures and weak precipitation rates. The strong insulation
allows the formation of mesoscale circulation (like sea breezes) and photochemical production (case A in Fig. 8.3). Nevertheless, a considerable number of photochemical $O_3$ episodes also develops under the influence of a thermal surface low-
pressure system, common over the Peninsula during summer, and with a strong effect on the development of a typical sea breeze (case B in Fig. 8.3) (Borrego et al.
1994; Barros and Borrego 1995). In the winter cyclone families either travel from
the West to the East or the Azores High expands eastwards.

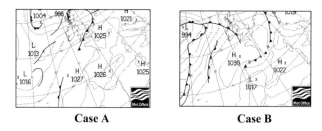

Case A                           Case B

**Fig. 8.3.** Typical meteorological summer conditions in the Iberian Peninsula (URL 8.2)

## 8.2.2 Synoptic conditions leading to the most severe photochemical pollution episodes

Air quality episodes are usually observed in a domain of 1000 km or less in which
mesoscale dynamics play a significant role. The severity of $O_3$ episodes may be
estimated with numerical models by considering surface and upper air meteorological data. High local $O_3$ concentrations observed in episodes are often influenced by mesoscale circulations embedded in large, stagnant synoptic systems.
Under a stagnant high-pressure synoptic system, these local flows become a major
mechanism for dispersing, mixing and transporting air pollutants as was the case
for $O_3$ episodes in Athens (e.g., 25-26/8/94 and 14-15/9/94 – Kalabokas and
Bartzis 1998), Barcelona (e.g., 3-5/8/90 – Toll and Baldasano 2000), Madrid (13-

17/7/99 – San José and Salas 2000), Milan and Brescia (3-5/6/96, 1-6/5/98 – Gabusi et al. 2001; Silibello et al. 2001).

Severe pollution episodes over the Athens basin have been mainly associated with the weak southerly flows and weak sea breeze as well as with almost calm conditions. Kallos et al. (1993) showed that the worst air pollution episodes in Athens occur during days with critical balance between synoptic and mesoscale circulations and/or during days with warm advection in the lower troposphere. According to the results of Kassomenos et al. (1998b), the most favourable synoptic pattern for the accumulation of high concentrations over the Athens basin is the open anticyclonic circulation and for the occurrence of extreme events the closed anticyclone. In association with mesoscale patterns the most severe pollution problem in the Metropolitan area of Athens is the high concentration of $O_3$ in the warm period. The problem is also substantial in the cold period (Kassomenos et al. 1998b).

Although Italian cities are located at geographic positions characterised by different climate and circulation features, the present understanding of photochemical pollution episodes is strongly biased towards the northern part of the country, and mainly focused on the Po Valley characterised by high industrial, urban and traffic emissions. Many experimental research programmes and modelling studies have been carried out over this area, e.g. EUROTRAC-2 subprojects SATURN and LOOP (Kruger et al. 2000; Neftel et al. 2000; Silibello et al. 2000). In 1998 the PIPAPO field campaign investigated the $O_3$ production in the Po valley. The field campaign covered May and June. Two major episodes were observed during the periods 12-13/5 and 3-6/6/98 (c.f. Fig. 8.4). During the first episode the circulation was associated with weather type 5, while the second episode was characterised by weather classified as type 1 for the Italian peninsula (c.f. Fig. 8.2).

A preceding relevant episode (5-7/6/96) was reported as one of the most widespread in Western Europe, characterised by an extended high pressure system with

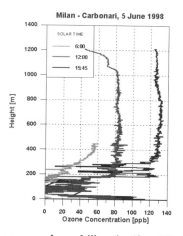

**Fig. 8.4.** $O_3$ vertical profiles measured over Milan city (from Finzi & 1999)

temperatures reaching 30°C and high $O_3$ levels covering Benelux and large portions of France, England, Germany and Denmark (de Leew et al. 1998; Silibello et al. 2001). Values exceeding the regional health protection threshold (110 µg m$^{-3}$ for 8 hours average) were recorded by 35 out of 48 stations in the Lombardia network. The synoptic weather conditions during the episode were characterised by a wide high pressure system covering Central and southern Europe at sea level, while at higher levels (700-500 hPa) the presence of a closed geopotential minimum, centred on southern Mediterranean sea, influenced southern Italy. The observed geopotential field can be associated to class 8 (Fig. 8.2). Other experiment in Milan urban area during July 2000 detected a typical summer high polluted episode (high $O_3$ and $NO_2$ levels, strong photochemical potential) (Finzi et al. 2000).

All the cited Italian photochemical episodes that have been analysed concern summer high pressure conditions, when subsidence and weak winds favour pollutant accumulation in the lower atmospheric layers. In the sub alpine region winter $O_3$ peaks can be observed during northerly föhn wind periods (weather type 6, or 9 in North-western regions). In these conditions $O_3$ concentrations are due to transport from upper troposphere or stratosphere, if stratopause breaking occurs. These episodes can be relevant at mountain foot and normally originate concentrations up to 100 µg/m$^3$ without carrying exceedances of air quality standards.

A severe pollution episode was observed over the Basque area during 14-16/6/96. During the episode a blocking anticyclone over Ireland on 13/6 moved eastwards and remained over the area until 16/6. A study of the $O_3$ episodes over the Basque area during the period 1995–1998, based on the available measurements of $O_3$ and the associated meteorology, show that the most important and persistent episodes were coincident with the onset of the east and North-easterly winds forced by the European High (Environment and Systems 1999; Gangoiti et al. 2002).

In the winter, pollution episodes have been also observed in the Iberian Peninsula. An example of a winter episode associated with synoptic conditions is the case of Madrid in January 1992. During the period 15-20/1/92 almost the whole region was under the influence of a high-pressure system centred over the British Isles. This situation produced very poor ventilation conditions over the Madrid area, increasing the air pollutant concentrations in the area (Pujadas et al. 2000). An $O_3$ episode which is characterised as "exceptional" occurring in the same area is that described by SanJosé et al. (2001). Exceptional $O_3$ episodes are considered those occurring at "non-expected" periods and they cannot be easily attributed purely to anthropogenic activities. Such an episode appeared in the night between 2h00-6h00 on 29/4/00 over Madrid and was attributed to $O_3$ intrusion from elevated layers that might have brought significant amounts of $O_3$ from the stratosphere as well as to the transfer of $O_3$ generated the day before by the prevailing wind.

## 8.3 Scale interactions and ozone formation

Evidently, apart from the synoptic conditions, mesoscale circulations play an important role in the formation and trapping of $O_3$ in certain areas. Any change in the pressure patterns associated with the two summer synoptic systems described in section 8.2 can significantly alter the balance among the mesoscale processes prevailing over the Mediterranean Basin, as well as determine whether the thermally induced local flows may develop or will be overruled by synoptic processes.

The surface properties and the presence of mountains and sea in the area of South Europe lead to the development of strong sea breezes, upslope winds or combinations of the two, depending on the mountain/coast orientation. Understanding the interaction of various scales (from street canyon geometry to the continental scale) may be extremely important for determining photochemical pollution episodes. As an example, the complex air flow configuration in a valley affects significantly the temporal and spatial variation of pollutant concentrations in both the valley itself (where usually cities are located) and its outskirts, i.e. the mountains around.

It is therefore important to deal with questions such as

1. Is there sufficient evidence from experimental campaigns or modelling studies to prove the importance of scale interactions regarding $O_3$ formation in particular southern European cities?
2. How does the background concentration of precursors in North and Central Europe affect the $O_3$ formation in South Europe?

### 8.3.1 Evidence from experimental campaigns and modelling studies

The complex airflow configuration in a valley affects significantly the temporal and spatial variation of pollutant concentrations in both the valley itself and the surrounding mountains. In places where the topography may limit or block surface winds under anticyclonic conditions, pollutants can be trapped within confined or semi-confined airsheds and serious pollution episodes can develop. Under such mesoscale conditions, an important aspect relating to atmospheric chemistry is that air mass aging can go on for days within the valleys, under oscillatory motions of limited amplitude, while new emissions are continuously being added to the air mass. These locally driven thermal circulations become weaker during the winter. Effects of mesoscale circulations on photochemical pollution and generally local air quality have been revealed with several experiments conducted in southern European cities, namely, LisbEx-96 (Lisbon, July 1996); the PIPAPO experiment (North Italy, May-June 1998); experiment in Province of Bolzano (July 1999); experiment in Milan (July 2000); ESCOMPTE campaign (South France, summer 2001); continuous measurements in Athens.

Cities in the Province of Bolzano in northern Italy experience smog episodes both in winter and summer exceeding the European air quality threshold values both for $NO_x$ in winter, and for $O_3$ in summer. A typical summer photochemical episode is due to prevailing valley winds during day and mountain winds during night. The local circulation in the area is found to be well uncoupled from the synoptic flow indicating that the wind flow in the valleys is generated mainly by local and regional thermal characteristics. Typical summer photochemical peak $O_3$ concentrations have been measured in the mountains surrounding the valleys where the city cores are located (Dosio et al. 2001). Due to the complexity of the flow in the area of the Po valley a large number of simulations has been performed mainly based on the results of the PIPAPO experimental campaign in order to validate different photochemical transport models. The simulation domains are characterised by complex terrain and high anthropogenic emissions in the area. The interaction of different scales was investigated providing local domain (Milano and Brescia metropolitan areas) simulations of the episode observed between 1 and 5/6/98 by means of nested procedures (Finzi & 2001). Urban and rural air quality impact simulations have been performed with CALGRID-FCM and STEMII-FCM using different chemical schemes (Fig. 8.5). Cities in the area of Catalonia also exhibit mountain-valley winds. Simulated surface wind fields calculated by the model MM5 at 12 UTC on a day (29/5/00) with a typical synoptic situation known as western advection showed the important role of the Pyrenees mountain range in the establishment of the regional circulations in Catalonia (Baldasano & 2001).

The area around Athens is subjected to a sea-breeze circulation. Over the period 17/7 to 6/8/97 each day was classified according to the wind pattern: (i) days with a strong northerly wind, suppressing the sea breeze (Etesian day); (ii) days with a fully developed sea breeze circulation; and (iii) days with moderate northerly winds and a weak sea breeze cell near the coast (Colbeck & 1999). Sea breeze circulation and its effects on $O_3$ concentration have been also observed in the area of Barcelona. During the early hours of the day, until 9 LST, $O_3$ is depleted where NO is being emitted, mainly by road traffic both in the urban areas and at the main communication routes, as well as at a point source (glass manufacturer) located in South Barcelona (Zona Franca). Meanwhile, the land breeze transports this low $O_3$

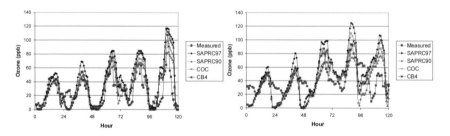

**Fig. 8.5.** Monitored and simulated $O_3$ concentration time series for Milan (urban site) and Erba (rural site) (Finzi & 2001)

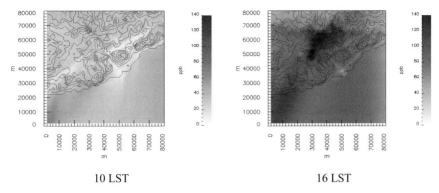

10 LST                                                  16 LST

**Fig. 8.6.** Horizontal cross-sections of simulated $O_3$ in Barcelona at 10LST and 16LST

concentration air mass out over the Mediterranean Sea. The sea breeze begins flowing inland between 10 and 12 LST and during the subsequent hours it is channelled inland through the Llobregat and Besòs valleys into the Vallès depression. This sea breeze causes the penetration of air masses loaded with $O_3$ from the Mediterranean Sea, which enters the domain through the lateral boundary conditions. The application of the meteorology/chemistry modelling system MEMO/MARS showed that the sea breeze flow is reinforced on the southern slopes of the coastal mountain range with the anabatic winds and does not penetrate beyond Collserola and Garraf until 16 LST (c.f. Fig. 8.6) (Baldasano ♄ 2001)

A characteristic $O_3$ episode showing the relation between rural and urban photochemical pollution is that of Athens on 25-26/8/94 and 14-15/9/94 as was investigated at an air pollution and meteorological station downwind of the urban area during sea-breeze periods (Kalabokas and Bartzis 1998) or upwind of the urban area when NE winds prevail. High $O_3$ was recorded at the measuring site due to the fact that it is downwind the urban area during the development of sea breeze. The maximum hourly average $O_3$ mixing ratio from the urban sector was about 50ppb higher than in the rural sector. In addition, during morning hours the average $O_3$ was increasing (mean wind speed of approximately 3-4m/s). This was attributed to the daily photochemistry as well as to the enhancement with $O_3$ produced the previous day and transported by convection from the upper layers or just produced elsewhere and advected to the area.

Results from the LisbEx-96 experiment in the area of Lisbon showed that the $O_3$ concentration data collected at several monitoring stations (e.g. Fig. 8.7) fits well with the expected pattern for this region. Specifically, typical behaviour can be identified for the coastal rural (M. Velho) and urban (Século) stations. The data collected on these stations is characterised by relatively low data dispersion. At the same time, the analysis for central tendency demonstrates lower values for the urban station (because of $O_3$ consumption during the day associated with traffic emissions) and higher values for the rural coastal station due to the low deposition rate over the ocean (Borrego & 2001).

More information and evidence on the aspect of the relation of different scale phenomena and $O_3$ formation is expected to be found with the analysis of the measurements taken during the ESCOMPTE experimental campaign performed in the greater area of Marseille in summer 2001.

At street scale, where airflow highly depends, apart from street characteristics, on the coupling with the layer above the street canyon, the use of a fast chemistry module ($NO-NO_2-O_3$ fast cycles) predicts high $NO_x$ levels on the leeward side of a narrow street canyon, while $O_3$ is depleted within the street canyon (Sahm et al. 2000) (Fig. 8.8).

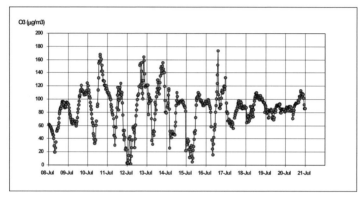

**Fig. 8.7.** Distribution of $O_3$ concentration at Tires station (LisbEx-96)

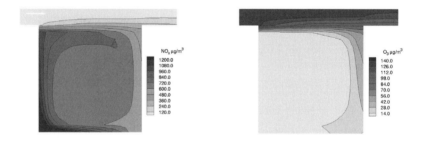

**Fig. 8.8.** $NO_x$ (left) and $O_3$ (right) fields within a street canyon as predicted with the use of fast chemistry

### 8.3.2 Effect of background concentration of precursors in North and Central Europe

At the larger scales, strong up-slope winds formed e.g. in the Alps and Atlas mountains, can inject aged air masses directly into the mid-troposphere and into the upper troposphere, where they can participate in long-range transport processes within southern and Central Europe, and at the continental-global scales, respectively. Therefore, the formation of $O_3$ varies greatly across Europe. There are indications that without anthropogenic emissions in Central and northern Europe, the background values in South Europe would drop considerably. During summer, an hemispherical $O_3$ background of approximately 40-50 ppb has been estimated at the South Europe latitude and it has been ascribed to anthropogenic sources.

## 8.4 Regional and local emissions of ozone precursors

In general, $O_3$ production is associated with the regional and local emissions of $NO_x$ and VOCs. It is therefore important to address issues like:

1. What is the contribution of regional and local emissions to the photochemical pollution in southern European cities?
2. What is the relation between regional and local emissions of $O_3$ precursors for the southern European cities?
3. How important is the role of biogenic VOCs as precursors of $O_3$?
4. Since agriculture is one of the main characteristics of southern European countries, how important are $NO_x$ emissions from fertilised soils?
5. Are there any "hot spot" areas and what are their characteristics?

$NO_x$ emission sources in cities are traffic and industry, while $NO_x$ may to some extent be emitted from fertilised soils. VOCs have many sources: traffic including the evaporation of fuel, industry, human activities and also biogenic sources. In case of pollutant episodes such as $O_3$ episodes during summer periods or $NO_x$ episodes during winter atmospheric stability periods – i.e. atmospheric high pressure – the need to take actions in real-time such as reduction of traffic flow or increase of use of public transport is essential for the success of those emission reduction strategic plans to avoid the exceedance of air quality standards.

### 8.4.1 Contribution of regional and local emissions

Model simulations have shown that for sensible reduction of the $O_3$ levels in an urban area, emission reductions are needed in both local and regional scales. Numerical studies have demonstrated that urban VOC control is effective in reducing $O_3$ primarily on the local or urban scale, while urban $NO_x$ control may cause an increase in peak concentrations on the local scale but is effective in reducing $O_3$ on the regional scale (Fig. 8.9, Table 8.1) (Moussiopoulos et al. 1998).

The vehicles velocity also affects $O_3$ precursors concentration; vehicles at low speeds emit larger amounts of VOCs and CO, while high $NO_x$ is also observed in major roads where the vehicles velocity increases (Toll and Baldasano 1999).

For the Lombardia Region, the impact of EU Directive on road traffic emissions scheduled for 2005 has been analysed by means of a photochemical modelling system (Volta and Finzi 1999; Veraldi et al. 2000; Finzi et al. 2000; Gabusi et al. 2001). The Present EU Directives on the road traffic emission will yield reductions of about 45% on $NO_x$ and CO and 35% on NMVOC in respect to 1996. The higher reduction percentages will result in urban areas. As a consequence of the emission sectors distribution all over the domain (including area and point sources), it comes out an estimated total reduction amount of about 20% for $NO_x$ and CO and 15% for NMVOC (Fig. 8.10).

Urban air quality impact simulations have been performed both for the reference emission field and for the 2005 updated estimate; both modelling system runs have been fed by the base case meteorological fields, in order to point out the environmental impact due to the assumed EU pollution abatement strategies. The simulated $O_3$ fields denote an increasing $O_3$ concentration (more than 10%) in the main metropolitan areas compared to base-case values (Fig. 8.11), instead of the reduction expected by legislators. The impact on PAN fields also presents similar features. The negative trend is consequent to the fact that, although the densely inhabited Lombardia region is characterised by a VOC-limited atmospheric chemistry system, the EU traffic emissions abatement strategies are mainly focused on measures leading to a prevailing global reduction on $NO_x$ emissions. However, the study advises the need of a deeper understanding of the phenomenon, both from the available measured data and the scientific knowledge points of view.

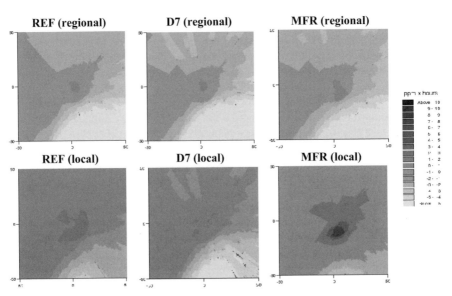

**Fig. 8.9.** The impact of the regional and local emission reduction scenarios on the AOT60 levels for various combinations of $NO_x$ and VOC reductions in the Greater Athens area

**Table 8.1.** Emission changes to the base case (1990 situation) for three scenarios

|  | REF | | D7 | | MFR | |
|---|---|---|---|---|---|---|
|  | NO$_x$ | VOC | NO$_x$ | VOC | NO$_x$ | VOC |
| *Greece (countrwide)* | -17% | -32% | -18% | -41% | -47% | -43% |
| *Athens* | -47% | -51% | -47% | -65% | -65% | -69% |

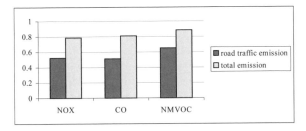

**Fig. 8.10.** Emission ratio in Lombardia between the 2005 and the base-case scenario 1996

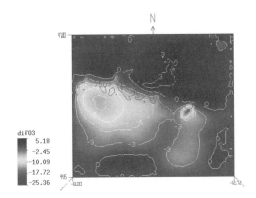

**Fig. 8.11.** O$_3$ variation in Lombardia (ppb) for the 2005 scenario with respect to the base case (5 p.m.)

## 8.4.2 Relation between regional and local emissions of ozone precursors

Within the framework of SATURN several emission models have been developed and improved. Many of them have been utilised for investigating the relation between regional and local emissions of O$_3$ precursors in South European areas. For example, for the area of Barcelona, it has been shown that highest CO and VOCs emissions take place in the city and are closely related to traffic (emissions of those pollutants are higher at the low speeds as occurring in cities). Some important highway junctions in the road network of the area also show important emissions of these pollutants. High emissions of $NO_x$ are found all over the road network, but the city of Barcelona does show the maximum magnitude as the rest of the roads since emissions of this pollutant decrease with decreasing speeds, compensating thus the higher number of vehicles in the city. Hourly distributions of VOC and $NO_x$ have indicated that highest emissions from road traffic occurred at 12 and 19 LST, when the volume of traffic was at its highest.

The combination of mesoscale circulations and local emissions strongly influence the production and spatial distribution of $O_3$ in the region. It has been observed how $O_3$ is formed over vegetated areas, where VOCs are emitted, due to the inland advection of $NO_x$ from the city with the sea breeze flow. Orographic injections from the surrounding mountains cause the formation of elevated $O_3$ layers. These elevated layers are dispersed according to the synoptic-scale flow, which is responsible for the orientation of the surface-height decoupled layers (Baldasano & 2001).

Model simulations for the city of Madrid have shown that $O_3$ concentrations are strongly related to the spatial distribution of primary emissions of $NO_x$ and VOCs and in spite of a strong increase in $O_3$ concentrations over the city domain when city traffic is not considered, the overall increase in $O_3$ is much larger when surrounding city traffic emissions are subtracted from the total emissions. Hence, the spatial distribution of traffic emissions is important for urban $O_3$ management policies (San José & 2001). In particular, the case investigated thoroughly was that of 13-17 July 1999 and three different scenarios were defined: A) Full Scenario: All emissions in the model domain during 120 hours of simulation. B) The emissions due to the traffic of Madrid Municipality are subtracted from the total emissions in those cells corresponding to the Madrid Municipality Area. C) The emissions due to the traffic of the surrounding areas of the Madrid Municipality Area are subtracted from the total emissions. Fig. 8.12a shows the $O_3$ air concentrations spatially averaged for the Madrid domain and when the scenario C and scenario B are subtracted (C-B=(C-A)-(B-A)) for the period starting on 13/7/99. Fig. 8.12b shows the $O_3$ air concentrations temporally averaged for the Madrid domain for the C-B scenario. The results show that surrounding areas get $O_3$ concentration increases about 36% in and central areas of the domain obtain $O_3$ decreases up to 18%.

The most recent Portuguese atmospheric emissions inventory is based on the CORINAIR methodology and reports emissions at national level for 1995. In order to obtain the spatial resolution required by air quality models, the area sources emission have been disaggregated, using statistical indicators. The results of this "top-down" approach are presented in Fig. 8.13 for the regional and urban Lisbon domain (Monteiro et al. 2002). Quantitatively, Lisbon city is responsible for 17% of $NO_x$ emissions and 12% of VOC emissions, comparing to the total regional emissions (Fig. 8.14).

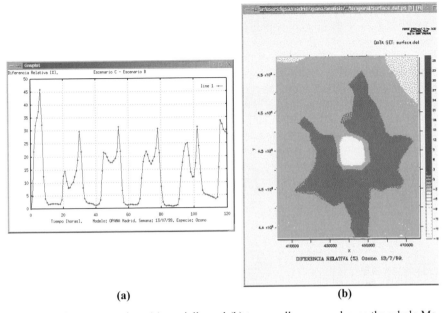

**(a)**                                                    **(b)**

**Fig. 8.12.** $O_3$ concentrations (a) spatially and (b) temporally averaged over the whole Madrid domain for 13-17 July 1999. Scenario C-B (from San José & 2001)

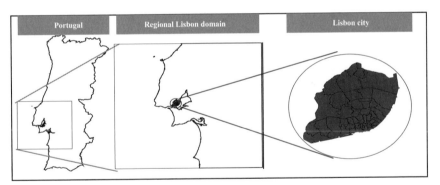

**Fig. 8.13.** Regional and Lisbon City domains

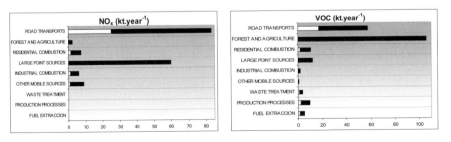

**Fig. 8.14.** Emissions comparison from Regional Lisbon domain and Lisbon City

The emission processor POEM-PM (Finzi et al. 2002) has been specifically designed to produce present and alternative emission field estimates by means of an
integrated *top-down* and *bottom-up* approach. POEM-PM can be applied in particular to the Italian CORINAIR data and considers diffuse and main point sources
coming from different activity sectors in the Po Valley. Thanks to its technology
and fuel-oriented formulation, the emission processor can be used to provide scenarios consistent with new fuel trades and pollutant abatement technologies.
Model outputs are the results of all possible combinations of three steps: the spatial disaggregation, the time modulation, the VOC and PM splitting. The road
transport, agriculture and biogenic emissions are estimated by means of the *bottom-up* approach implemented by the model up to 1996 and 1998. The emission
fields due to the other sectors are computed by means of the *top-down* approach
starting from the CORINAIR90 data set, updated by the last available Italian
CORINAIR report (1994). Yearly emissions computed by POEMPM for the
Lombardia Region domain (240x232 km$^2$) are summarised in Table 8.2.

**Table 8.2.** Yearly emissions [Mg] within the domains

|  | Lombardia Region (240×232 km2) | | Milan Metropolitan area (100×100 km2) | |
|---|---|---|---|---|
|  | NO$_x$ | VOC | NO$_x$ | VOC |
| *traffic* | 268 450 | 167 060 | 44% | 66% |
| *industry* | 49 372 | 31 853 | 58% | 76% |
| *resid comb* | 21 910 | 4 190 | 60% | 41% |
| *solvent use* | 0 | 167 000 | 0% | 55% |

### 8.4.3 Importance of the role of biogenic VOCs

Regarding the biogenic VOCs, their highest emission is associated with highest intensity of solar radiation and high air temperature as was observed e.g. for the
Greater Barcelona Area. However, monoterpenes are the most emitted BVOC during the whole day, whilst when isoprene is greatest (i.e. between 14 and 16 LST)
monoterpenes are lower. A sensitivity analysis to ascertain the importance of isoprene in O$_3$ abatement strategies has shown that its emissions do not affect the
load to any great extent in respect of long-term average values (Simpson 1995).

In Portugal, sensitivity tests for VOC biogenic emissions with a photochemical
model allowed verifying a less significant contribution of these emissions to the
photochemical O$_3$ production. In Fig. 8.15, the differences between a baseline case
and a simulation without VOC biogenic emissions are presented.

A quantitative analysis (Table 8.3) shows that these differences are not significant when compared to the contribution of others emissions sources, like road transport and large point sources emissions.

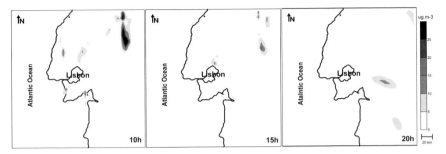

**Fig. 8.15.** Spatial $O_3$ concentration differences between baseline simulation and without biogenic emissions

**Table 8.3.** Cell reduction percentage, on $O_3$ production, considering three scenarios, without: biogenic emissions, industrial point sources and road transport emissions

| % $O_3$ cell reduction | Biogenic | Industrial point sources | Road transport |
|---|---|---|---|
| *Maximum % of $O_3$ reduction* | 0.12 | 0.56 | 0.76 |
| *Median % of $O_3$ reduction* | 0.002 | 0.027 | 0.062 |
| *N° of cells with $O_3$ reduction* | 869 | 1283 | 1846 |

## 8.4.4 Importance of $NO_x$ emissions from fertilised soils

Sensitivity modelling tests were performed in order to evaluate the contribution of $NO_x$ emissions from agriculture and nature in $O_3$ production. Slightly lower $O_3$ concentrations are obtained in the case that $NO_x$ agriculture emissions are neglected (Table 8.4).

**Table 8.4.** Cell reduction percentage, on $O_3$ production, obtained with a simulation without considering agriculture emissions

| % cell reduction | 10h | 11h | 12h | 13h | 14h | 15h | 16h | 18h | 20h |
|---|---|---|---|---|---|---|---|---|---|
| *Maximum %* | 22.2 | 32.3 | 20.7 | 11.1 | 7.6 | 6.8 | 5.2 | 8.8 | 11.2 |
| *Median %* | 0.38 | 0.74 | 0.64 | 0.44 | 0.32 | 0.35 | 0.37 | 0.42 | 0.86 |
| *N° of cells* | 1506 | 1603 | 1686 | 1661 | 1660 | 1272 | 1117 | 1108 | 1478 |

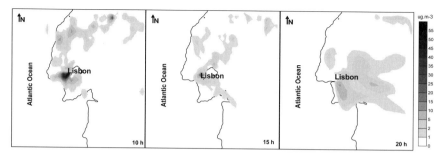

**Fig. 8.16.** Spatial $O_3$ concentration differences between baseline simulation, without considering emissions from agriculture

The spatial differences in $O_3$ production between simulations considering and not considering agriculture emissions are presented in Fig. 8.16. A significant area with $O_3$ reduction, but with low values, can be observed. In fact, agriculture is not a main polluting activity, when compared with other emissions sources, like road transport and large industrial point sources.

### 8.4.5 "Hot spot" areas

The $O_3$ concentrations fields obtained for the two main Portuguese urban areas shows that some $O_3$ "hot spot" areas exist beyond the two large cities of Porto (Borrego et al. 1998) and Lisbon (Borrego et al. 2000). In Fig. 8.17a and b, it is possible to identify two critical coastal zones South-East of the two cities, considering synoptic conditions of a typical summer day (corresponding to the most critical conditions for $O_3$ production).

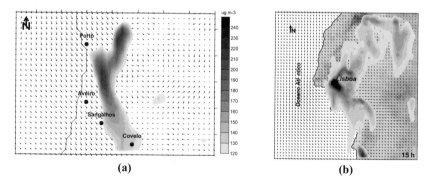

**Fig. 8.17.** $O_3$ concentration field on a typical summer day for **(a)** Porto and **(b)** Lisbon areas

The $O_3$ concentration fields simulated by photochemical models for North Italy in different episodes show two "hot spot" areas located to the North and South of Milan metropolitan area (Fig. 8.18a). This remarkable phenomenon can be explained by the meteorological condition prevailing during the photochemical episodes: the dominant wind directions are biased by mountain-valley breeze regimes. The Milan plume is, therefore, driven both North-eastwards forced by Alps, in particular, by Valtellina Valley, and southwards forced by Apennines (Fig. 8.18b).

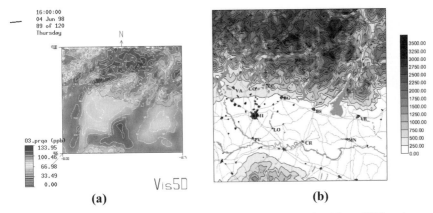

(a)                                    (b)

**Fig. 8.18. (a)** $O_3$ concentration field for Lombardia Region Domain, 4 June 1998 at 4 p.m. **(b)** Lombardia Region domain orography

## 8.5 Indicators of photochemical pollution

The formulation of an $O_3$ control strategy requires the knowledge of the emission control derivatives $d[O_3]/[dE_{NOx}]$ and $d[O_3]/[dE_{VOC}]$ associated with the response of $O_3$ to changes in emissions of $NO_x$ and VOCs. One way of approaching this problem is to look for indicator compounds, which maintain memory of whether $O_3$ was formed under $NO_x$ or VOC sensitivity conditions. Reactions forming $O_3$ also yield oxidation products that can be used to determine whether $O_3$ was produced under $NO_x$- or VOC-sensitive conditions. The VOC-sensitive regime (or $NO_x$-saturated) is characterised by an increase in $O_3$ when there is an increase in VOC or decrease in $O_3$ when there is further increase in $NO_x$. On the other hand, the $NO_x$-sensitive regime is marked by increase in $O_3$ for an increase in $NO_x$. The species often used as indicators are (Sillman 1995):

- the ratio $H_2O_2/HNO_3$: VOC-limited conditions are associated with low values of the ratio, while $NO_x$-limited points are associated with high values of this ratio;

- the ratio $O_3/NO_z$ and other variants computed by subtracting background concentrations or by replacing $NO_z$ with $HNO_3$ or $NO_y$: $DO_3/DNO_y$ is the number of molecules of $O_3$ formed per $NO_x$ molecule emitted. $DO_3/DNO_y$ takes into account partial reaction of emitted $NO_x$ as it represents the $O_3$ formed per $NO_x$ consumed. High values of $O_3/NO_z$ are associated with $NO_x$-limited conditions and low values associated with VOC-limited conditions. $NO_y$ is the sum of $NO_x$ and $NO_z$ with $NO_z$ representing the sum of oxidation products of $NO_x$, i.e. $NO_z = HNO_3$ + organic nitrates + inorganic nitrates.

In reality, it is not easy to identify accurately the threshold line for which the photochemical regime changes from being VOC to $NO_x$ sensitive; for this reason a broad transition area should be considered (Sillman 1999). The transfer from the one regime to the other is closely related with the age of the air mass. Typically, a freshly emitted plume of polluted air is more likely to be characterised by VOC sensitivity and it evolves to $NO_x$ sensitive for "older" air masses. The speed of conversion from VOC- to $NO_x$-sensitive chemistry depends on how rapidly the $NO_x$ in the air mass reacts away. Photochemically aged air is also found after multiday transport. VOC-sensitive chemistry is most likely to occur in central locations in urban areas, as was shown for Los Angeles by field experiments (e.g., Lawson 1990) or numerical model studies (e.g., Winner et al. 1995). These studies also showed that $NO_x$-sensitive chemistry is mainly found at downwind locations. Similar behaviour has also been reported in European cities such as Milan (Prevot et al. 1997; Silibello et al. 2000).

$NO_x$-VOC chemistry is related to the relative size of the sources of odd hydrogen radicals and $NO_x$. The rate of $O_3$ production is determined by the rate of $NO_x$ removal and the $O_3$ production efficiency. In the VOC-sensitive regime, the rate of removal of $NO_x$ is limited by the availability of radicals. Hence, the rate of chemical processing of $NO_x$ does not increase with increasing $NO_x$. At the same time, increased $NO_x$ is associated with lower $O_3$ production efficiency. On the other hand, in the $NO_x$-sensitive regime an increase in $NO_x$ concentration is associated with increased removal of $NO_x$ and therefore with increased $O_3$ production (Sillman 1999).

The highest $O_3$ concentrations are typically found in urban air masses as they are moved downwind of the city centre. Peak $O_3$ concentrations usually occur 50-100km from the city centre or even further downwind especially in coastal regions. Hence, both $NO_x$ and VOC-sensitivity conditions are closely related and characterise an $O_3$ episode. Therefore, in order to address an $O_3$ episode evolving in southern European urban areas it is important to ask:

1. What are the main indicator compounds used so far in southern European studies?
2. Which are the dominant sensitivity regimes in particular southern European cities?
3. Is there a use of Observation-Based Model (OBM) for reconstructing the dependence on emissions from a sequence of present time-frame observations?

## 8.5.1 Main indicator compounds and $NO_x$ and VOC sensitivity

The indicator species approach has the advantage of requiring only a point measurement of a pair of species such as $H_2O_2$ and $NO_z$. This approach is used to maintain a record of whether $O_3$ was formed under $NO_x$ or VOC sensitivity conditions.

Model studies (Milford et al. 1994; Sillman 1995) have pointed out that it is possible to predict $O_3$-$NO_x$-VOC sensitivity (and consequently the more effective emission control strategy) according to values assumed by specific species and species ratios (c.f. Table 8.5). Specifically, when the models predict that $O_3$ is primarily sensitive to $NO_x$, they also predict high values for certain species ratios, namely $O_3/NO_y$, $O_3/NO_z$, $H_2O_2/HNO_3$, $H_2O_2/NO_z$, etc. On the other hand, when models predict that $O_3$ is primarily sensitive to VOC, they also predict low values for the same ratios. Therefore, measured values of these ratios can be used to test model predictions and evaluate $O_3$-$NO_x$-VOC sensitivity directly (Neftel et al. 2002).

The application of widely used models to the Milan plume under the framework of LOOP (some of these models also used under the framework of SATURN) resulted to the same transition value for $H_2O_2/HNO_3$ namely 0.2. Although this set of simulations is limited it may be suggested that a stable transition value for $H_2O_2/HNO_3$ exists which is of the order of 0.2. Additional simulations have to be carried out to study the influence of different chemical mechanisms and of boundary conditions on the transition value of $H_2O_2/HNO_3$. On the other hand, $NO_y$ is an indicator with the limitation that it can only be applied to very simple situations with very clear emission conditions e.g. isolated sources, while no single transition value exists for this indicator. The results of Vogel et al. (1999) have shown that the transition value of $NO_y$ depends on ambient conditions such as temperature, humidity and radiation and on the emission conditions.

Results from the PIPAPO field campaign suggest that the Milan plume plays a key role shifting the generally $NO_x$ sensitive $O_3$ production regime towards a VOC sensitive regime. Further arguments to confirm this suggestion come from the use of the indicator concept. The geographical variation in VOC/$NO_x$ chemistry, influenced by different factors such as VOC/$NO_x$ emission ratio, VOC reactivity, photochemical ageing and dispersion, has been analysed by means of the indicators $NO_y$, $O_3/NO_z$, $HCHO/NO_y$, $H_2O_2/HNO_3$. The same experiment showed that the scale length of the $O_3$ production (i.e. the distance where a chemically "fresh" mixture is aged to $NO_z/NO_y > 0.75$) can be only a few tens of kilometres in Central and southern Europe under typical summer conditions (in URL 8.3).

In the framework of SATURN, the concentration fields estimated for the episode 3-5/6/96 in the Lombardia region have been analysed in order to point out the chemical regime of the area (Silibello et al. 2000). The indicator approach suggests that VOC control is more effective when the afternoon $NO_y$ and the $O_3/NO_z$ exceed 10-25 ppb and are lower than 6-11, respectively. Values within these ranges fall in the so-called transition regime. Predicted ground-level fields of $NO_y$ and $O_3/NO_z$ ratio on a summer afternoon are almost complementary evidencing the capability of the chosen indicators to actually separate the two chemical regimes (Fig. 8.19). A VOC sensitive area covers a large part of the Po Valley, where major urban areas are present, whilst a $NO_x$ sensitive one covers the remaining part of the domain including the Alps region. Transition from VOC to $NO_x$ sensitive $O_3$ production can occur at relatively high $NO_x$ concentrations.

In order to verify these results, two different emission scenarios have been used considering a uniform 35% reduction in the emission rates over the whole area respectively for anthropogenic VOC and $NO_x$. Figs. 8.20 and 8.21 show the predicted reduction in ground-level $O_3$ concentrations ($O_3^{base\ case}$ - $O_3^{scenario}$), for all grid-points of the domain and at 14 UTC 7 June, against $NO_y$, $O_3/NO_z$, $HCHO/NO_y$ and $H_2O_2/HNO_3$ indicators. The analysis of these plots and of literature threshold values for $NO_x$ sensitivity of the selected indicators (Table 8.5) confirms the capability of the indicator approach to characterise the chemical regimes within the studied area and its robustness. In fact predicted indicator threshold values fall in the proposed ranges for different $NO_x$-VOC sensitivity regions.

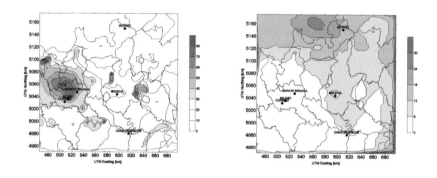

**Fig. 8.19.** Computed ground-level $NO_y$ concentration field (left) and $O_3/NO_z$ field (right) at 14 UTC on 7 June

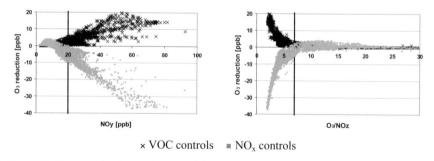

× VOC controls    ■ $NO_x$ controls

**Fig. 8.20.** Computed reduction in ground-level $O_3$ concentration field, at 14 UTC of 7 June, resulting from a 35 % reduction in the emission rate for anthropogenic VOC (black crosses) and from a 35 % reduction in the emission rate for $NO_x$ (grey squares) plotted against $NO_y$ (left) and $O_3/NO_z$ (right)

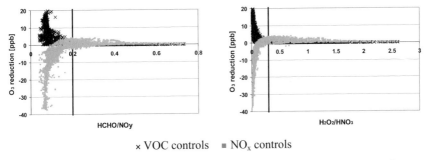

× VOC controls    ■ NO$_x$ controls

**Fig. 8.21.** Computed reduction in ground-level O$_3$ concentration field, at 14 UTC of 7 June, resulting from a 35 % reduction in the emission rate for anthropogenic VOC (black crosses) and from a 35 % reduction in the emission rate for NO$_x$ (grey squares), plotted against HCHO/NO$_y$ (left) and H$_2$O$_2$/HNO$_3$ (right)

**Table 8.5.** Threshold values for NO$_x$-sensitivity of afternoon concentrations or ratios (Silibello et al. 2000)

| Indicator | Threshold | Reference |
|---|---|---|
| | | Milford et al. 1994; |
| NO$_y$ | < 10 to 25 ppb | Sillman 1995 |
| O$_3$/NO$_z$ | > 6 to 11 | Sillman 1995 |
| HCHO/NO$_y$ | > 0.2 to 0.4 | Sillman 1995 |
| H$_2$O$_2$/HNO$_3$ | > 0.3 to 0.5 | Sillman 1995 |

Fig. 8.22 shows predicted changes in the ground-level afternoon O$_3$ concentration field for the two emission scenarios. Consistently with the indicator approach analysis, in the VOC sensitive region, characterised by NO$_y$ levels greater than 20 ppb (Fig. 8.22 - left), the control of NO$_x$ emissions (35% reduction in NO$_x$) leads to an increase of O$_3$ levels, up to 36 ppb, (Fig. 8.22 - left). The corresponding control of anthropogenic VOC determines a decrease of O$_3$ up to 16 ppb (Fig. 8.22 - right). In the complementary region (NO$_x$ sensitive), characterised by the presence of more aged air masses and lower anthropogenic sources, both emissions control strategies produce negligible effects on O$_3$ levels. Also, VOC sensitivity has been observed in the urban area of Lisbon and NO$_x$ sensitivity in neighbouring rural areas (Fig. 8.23).

The O$_3$ production has been investigated in the area of Crete (Greece) with the Pump and Probe (PP OH) LIDAR technique, a method based on the measurement of the atmospheric OH life time in very different air quality conditions. This is the first *in situ* technique able to directly characterise whether the probed volume is "NO$_x$ or VOC controlled" in terms of O$_3$ production. Furthermore, this technique is designed as a range resolved method. After a precise calibration of the sensitivity of the technique, first range resolved OH spectra were obtained. With the PP OH method, the determination of OH reactivity in different air pollution conditions, a new indicator is gained for the NOₓ/VOC control of the O₃ secondary production and thus will directly contribute to test and validate model results (Calpini $\mathscr{A}$ 2001).

The indicator concept opens the possibility of measuring the effects of emission reductions. In this way one could measure which areas of a certain domain would benefit from a NOₓ emission reduction and which areas would benefit from a VOC emission reduction. Up to now the measurement of indicators is not an easy task. So far aerosols have not been included in models used for indicator evaluation. The PIPAPO experimental campaign suggested that the aerosols are an important component of the indicator ratios. Therefore, future evaluation of indicator values should include both gas-phase and aerosol chemistry (Neftel et al. 2002).

It is rather difficult to recommend policies for the two photochemical regimes and to distinguish which regime is less harmful for the human health. Evaluations of O₃ policies that weigh impacts based on population exposure are more likely to favour VOC controls, while evaluations based on the impact of the O₃ concentration at low levels are more likely to favour NOₓ controls.

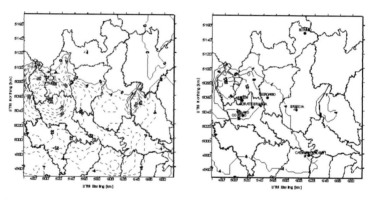

**Fig. 8.22.** Computed changes in ground-level O₃ concentration field for 35 % reduction in NOₓ emissions (left) and anthropogenic VOC (right). Positive values (dashed gradation) correspond to a decrease of O₃ concentrations due to adopted emission reduction strategy

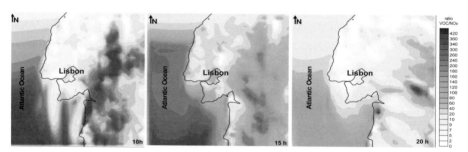

**Fig. 8.23.** Spatial VOC/NOₓ representation for a typical summer day simulation

### 8.5.2 Observation-Based Model

The Observation-Based Model (OBM) reconstructs the dependence on emissions from a sequence of present time-frame observations. Data requirements for OBM are more extensive than the indicator method; time series of observations of $O_3$, NO, CO, and VOCs are required. This method yields quantitative predictions on the response of $O_3$ to changes in $NO_x$ and specific VOCs.

Based on the findings of the PIPAPO experiment, the LOOP subproject suggested that

- Peroxide inhibition in plumes appears to be an indicator for VOC sensitivity. This inhibition needs to be discussed in comparison to deposition and possible vertical mixing.
- It is difficult to find situations where a plume is advected without additional emissions. Therefore, extracting information on OH concentrations from time dependence of $NO_x$ or VOCs is probably not possible.
- There are observations in source regions. Ratios of primary compounds (i.e. $CO/NO_x$, $VOC/NO_x$, $SO_2/NO_x$, ratios of VOC) can be compared with emission numbers to assess accuracy of inventory.
- For ground stations, $NO_z$ is preferable to $NO_y$ because of possible large effects from nearby $NO_x$ sources.
- If the vertical profiles show that the air is well-mixed, surface observation should be used for indicator ratios.

## 8.6 The role of aerosols

Aerosols are increasingly implicated in matters of health via their direct impact through respiration and through their secondary impact on oxidant levels. In order to evaluate their significance in these areas, an improved knowledge is needed of aerosol composition relative to emissions. How does composition change downwind of a source? One would expect evolution in size, composition and number density, but quantitative results are very limited. Aerosol characterisation is further hindered by our poor understanding of formation processes that contribute to the relative abundance of primary aerosols, which are formed via mechanical or combustion processes, and secondary aerosols, which are formed via gas-phase condensation.

Aerosols have been implicated in the concentration, lifetime, and chemical behaviour of sulphur, nitrogen and carbon compounds. They are also thought to provide reaction sites and to serve as carriers for many atmospheric trace gas species. Yet much work remains to define the gaseous species that lead to aerosol formation, and to identify key surface reactions under various conditions, e.g., polluted urban air versus relatively clean marine air.

Long-range transport of photochemical gaseous air pollutants and particulate matter has been studied extensively in Europe the last decades under the framework of several national and international efforts. Furthermore, available information on characteristics of photochemical pollutants/fine particles in southern Europe and their effects on air quality are also limited. There is a consistent pattern of geographical variability in Europe with lower concentrations of particulate matter in the far North and higher concentrations in the southern countries. This is due to natural emissions of unsaturated hydrocarbons (including isoprene) which are highly reactive and continuing high emissions of anthropogenic gaseous and aerosol pollutants in southern Europe. The Mediterranean is characterised by specific natural aerosol load namely sea spray and North African Desert dust. These natural particulate emissions are involved in heterogeneous reactions with anthropogenic gaseous pollutants and may modify the processes leading to gas to particle conversion. Important questions necessary to answer are:

1. How important are the nitrous acid (HONO) emissions in the southern European regions?
2. What is the influence of secondary and sea-salt aerosols as well as the Saharan dust on photochemical pollution?

### 8.6.1 The role of nitrous acid (HONO)

Nitrous acid, HONO, is a source of the most important daytime radical: the hydroxyl radical (OH). OH is one of the key species in photochemical cycles responsible for $O_3$ formation, which can lead to photochemical smog in polluted regions. Observations of HONO in the atmosphere show a typical diurnal cycle. Since the major loss process is photolysis, HONO builds up during the night. At sunrise it is photolysed into OH radicals and NO and its concentration drops. During the early morning hours HONO can be the most important source of OH in the polluted atmosphere. The concentration remains low during the day and rises again after sunset.

To understand the mechanisms that lead to this formation we need to know the sources of OH. There is ample evidence that HONO is formed from $NO_2$ in the presence of water in a heterogeneous reaction on both ground and airborne surfaces, i.e. aerosol particles. The reaction mechanism can be summarised as:

$$HONO + h\nu \rightarrow OH + NO \tag{8.1}$$

Laboratory measurements suggest that the HONO formation is first order in $NO_2$ and $H_2O$, but the exact mechanism is unknown. Recently, the formation of HONO on soot particles in the presence of $NO_2$ and perhaps water has been observed in the laboratory.

The sources of HONO are:

- *Homogeneous Chemistry*: HONO can be formed in the reverse reaction to (8.1)

$$NO + OH \rightarrow HONO \tag{8.2}$$

This reaction is insignificant during the night, because the OH concentrations are too low. Its significance during the day is currently unknown.

- *Heterogeneous Formation*: It is believed that HONO is mainly formed hetero-geneously on surfaces in the presence of water and $NO_2$. The reaction mechanism can be summarized as:

$$H_2O + 2\ NO_2 \rightarrow HNO_3 + HONO \tag{8.3}$$

Observations indicate that NO is not necessary in the formation of HONO. Unfortunately the HONO production rates determined in the laboratory are about two orders of magnitude lower that the ones observed in the atmosphere. Since it is not known on which surfaces HONO is produced in the atmosphere, a translation of laboratory data to the atmosphere might not be valid.

- *Formation on Soot*: Recently, the formation of HONO on soot particles in the presence of $NO_2$ and perhaps water has been observed in the laboratory.

- *Direct Emission*: Another source of HONO is direct emissions from combustion processes.

The most important sink of HONO is photolysis. The heterogeneous self-reaction of HONO has been investigated in the laboratory at high HONO concentrations, but it is unlikely that it plays a role in the atmosphere. HONO can also react with secondary and tertiary amines, forming carcinogenic nitrosamines.

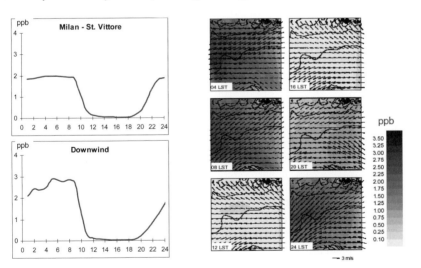

**Fig. 8.24.** HONO concentration at the centre of Milan and at a downwind location (left) and HONO concentration field and wind field predicted for Milan at different times (right) (Moussiopoulos et al. 2000)

Observations of HONO in the polluted atmosphere often show night time mixing ratios of more than 10 ppb. HONO has been found to be influenced by the location related to the city centre as well as by meteorological effects, i.e. the change from night-time to daytime chemistry as the predicted HONO for Milan area shows (Fig. 8.24).

The problem with HONO is that its concentration is strongly influenced by meteorological effects, i.e. the change from night time to daytime chemistry. A number of questions remain concerning HONO formation:

- How is HONO formed in the atmosphere, and which parameters influence its formation?
- On which surfaces is HONO formed in the atmosphere?
- What are the HONO production rates during the night and during the day?
- Is direct emission of HONO by combustion processes an important source?
- How significant is OH formation by HONO photolysis?

### 8.6.2 Influence of sea salt and secondary aerosols

Sea salt aerosol has been proposed to be a source of chlorine in the marine boundary layer. Molecular chlorine photolyses quite rapidly in the troposphere to form chlorine atoms. The chlorine atoms are highly reactive with organics, particularly natural hydrocarbons like isoprene. Monoterpenes are also quite reactive. The reaction rate of chlorine atoms with isoprene is five times faster that of OH radicals. At concentration levels of Cl and OH anticipated in marine air, the Cl radical will probably be very important in initiating early morning photochemical processes, while OH will be more dominant in hydrocarbon oxidation during the middle and later hours of the day. It has also been suggested that ammonium sulphate aerosols could promote the reactions of aliphatic alcohols with OH radicals. This process contributes to the production of secondary $NO_2$, ultimately leading to an enhancement in tropospheric $O_3$ levels.

Today in many major urban areas around the world, photochemical $O_3$ is the main concern among the many air pollution problems. The key reaction leading to $O_3$ formation in the troposphere is the photodissociation of $NO_2$ at $\lambda < 430$ nm:

$$NO_2 + h\nu \rightarrow NO + O \qquad (8.4)$$

The photolysis frequency $J(NO_2)$ depends on the solar spectral actinic flux which is dependent upon the local optical conditions. Radiative transfer models show that aerosols can either increase or decrease the actinic flux, depending on their scattering and absorption properties in the UV. Dickerson et al. (1997) were one of the first to consider that reductions in emissions of sulphur and hydrocarbons may yield unanticipated benefits in air quality. While sulphate and some organic aerosol particles scatter solar radiation back into space and can cool Earth's surface, they also change the actinic flux of ultraviolet (UV) radiation. Observations and numerical models show that UV-scattering particles in the boundary layer accelerate photochemical reactions and smog production, but UV-absorbing aerosols such as mineral dust and soot inhibit smog production. Dickerson et al. (1997) found for the eastern US a large increase in $O_3$ production due to scattering by non-absorbing aerosol. Castro et al. (2001) measured and calculated values of $J(NO_2)$ in Mexico City and report that the impact of aerosols can reduce the value of $J(NO_2)$ by 10-30%. Such reductions result in $O_3$ concentrations being diminished by several tens of ppb.

Bhugwant et al. (2000) monitored $O_3$ and black carbon (BC) aerosols at a tropical site. A comparison of $O_3$ and BC measurements points to some possible effects of heterogeneous interaction of $O_3$ and its precursors with BC particles. Model results show that adsorption of $O_3$ and its precursors onto BC aerosol particles could be one of the important steps determining $O_3$ concentration characteristics, especially in the absence of photochemistry during night-time.

Aerosol size distribution measurements performed in the free troposphere have been reported by de Reus et al. (2000). During one measurement flight a uniform aerosol layer was encountered between 2.5 and 5.5 km altitude, characterised by a relatively low Aitken mode particle number concentration and high concentrations of accumulation and coarse mode particles resulting in a relatively large aerosol surface area and mass. Five-day backward trajectories indicated that the aerosol in this layer was mineral dust originating from arid regions on the North African continent. The dust layer was associated with reduced $O_3$ mixing ratios. The best agreement between the observed and modelled $O_3$ concentrations was obtained when heterogeneous removal of $O_3$ and precursor gases on dust aerosol was taken into account. Heterogeneous $O_3$ loss was estimated at 4 ppb $O_3$ per day. He and Carmichael (1999) studied aerosol impact on $NO_2$ photolysis rates and $O_3$ production in the troposphere. They found that the presence of absorbing aerosols in the boundary layer inhibits near-ground $O_3$ formation and reduces ground level $O_3$ by up to 70% in polluted environments. The presence of strongly scattering aerosols may increase $O_3$ concentration in the lower boundary layer, but their effects vary with season, $NO_x$, nonmethane hydrocarbon emission (NMHC), and temperature. In the lower troposphere, $NO_2$ photolysis and $O_3$ production rates are most sensitive to urban aerosol, followed by rural, then desert, and finally, maritime aerosol. Gaffney et al. (2002) reported that a regional smoke episode in Arizona was accompanied by a decrease in UVB of factor of two and a decrease in $O_3$ and an increase in methyl chloride.

Most photochemical models currently in use do not take into account the solar radiation extinction by atmospheric aerosols. The implementation of solar radiation, gas phase and aerosol chemistry modules in model TAPOM (Transport and Air POllution Model) showed significant changes on $O_3$ concentrations. The main effects are seen on aerosol precursors such as nitric acid which is almost totally consumed when aerosol calculation is performed (Calpini & 2001).

Secondary particles directly converted from gaseous precursors are present in the fine particle size range whereas those formed from reactions with sea

salts/minerals such as NaCl and CaCO$_3$ are expected to be coarser as was measured in Pireus (Fig. 8.25).

The experimental activity in Milan area during the year 2000 has been mainly
focused on urban aerosol chemical composition and aerodynamic distribution
investigation. The results help to understand deeper which chemical multi-phase
modelling schemes have to be implemented in the mesoscale models. The experimental campaign was set up during July 2000; a typical summer high polluted episode (high O$_3$ and NO$_2$ levels, strong photochemical potential) was detected; aerosol analysis was combined with a complete characterisation of the primary and
photochemical pollutants both at ground level and in the vertical structure. The
sampling strategies have been devoted to characterise aerosol composition (ammonium, nitrate and sulphate, OC, EC, anions and cations) split into different
aerodynamic classes. Results mainly concern the urban aerosol mass closure determination and the relation between secondary aerosol formation and photochemical properties of the atmosphere. A preliminary evaluation of secondary
aerosol impact in the Po Valley has been performed by means the modelling system GAMES (Decanini et al. 2002), including the CALMET meteorological processor (Scire et al. 1999), the POEM-PM emission processor (Finzi et al. 2002) and
a recently developed photochemical model; such a model integrates the
CALGRID transport module of the FCM gas phase chemical mechanism, the
MAPS aerosol module and the UAM dry deposition module. GAMES has been
applied to simulate the O$_3$ episodes which occurred in early June 1998, in particular for PM modelling. Simulation results give good agreement with PM$_{10}$ and
PM$_{2.5}$ measurement data in the area, showing a close relation between O$_3$ accumulation and secondary aerosol dynamics.

Sea-salt aerosol in the presence of O$_3$ has been shown to produce Cl atoms in
heterogeneous photochemical reactions under laboratory conditions. Whether
chlorine can initiate oxidation of natural organics and can generate homogeneous
nucleation or condensable material that contributes to aerosol loadings needs to be
assessed. The nighttime reactions of O$_3$ and NO$_3$ radical can also result in
monoterpene reactions that contribute to aerosol mass.

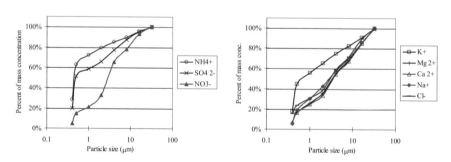

**Fig. 8.25.** Cumulative size distributions of ambient aerosols measured at Pireus (Colbeck
2000)

In order to study the effect of sea-salt aerosols on $O_3$ formation in the Mediterranean area field experiments were performed in Greece. In particular, a field experimental campaign in Athens showed that the wind regimes exhibit a distinct effect on the ambient aerosols in the urban area. Sea breeze circulation enhances the development of secondary aerosols which was clearly shown at the inland site measurement. Nitrous acid, hydrochloric acid and particulate nitrate, sulphate and ammonium increase during sea breeze days (Colbeck & 2000).

Other measurements of sea-salt aerosols were performed during summer and winter 2000 and 2001, respectively, at two sites in the Mediterranean: the Finokalia site on the island of Crete (Greece) and aboard the Aigaion (only in the summer 2000) located in the Eastern Mediterranean area between the Greek mainland and the island of Crete (Colbeck & 2001).

A number of questions remain concerning homogeneous aerosol formation by natural organics interacting with anthropogenic pollutants. For example, chlorine has been proposed as a potential oxidant in the troposphere because of its very high reactivity with a wide range of organics.

## 8.7 Real-time ozone forecast models

The European Community directive 92/72/EEC prescribes air quality standards in terms of threshold values for health protection, population information and warning. A proposal of an air quality alarm system has been designed, implementing two feedback loops (Fig. 8.26) by means of different forecast modelling methodologies (*non-stationary / non-linear grey box* and *neuro-fuzzy* predictors).

Air quality status forecast, given by the daily model, is supplied to the Authority in order to support decisions relevant to emission abatement strategies. It is also applied to inform the population by means of media, in order to prevent hazardous exposure. The designed system includes a second feedback improving operational effectiveness in the short term (hours). The information, provided both by the hourly predictor and by the on-line meteo-chemical network, allows the Air Quality Manager to monitor the current pollutant evolution. Any threshold value exceedance, not correctly forecast in advance, can be quickly recognised by the alarm system in order to apply short-term pollution control measures. The metropolitan area of Brescia has been considered as case study. The city is located in the Po Valley (northern Italy) and is characterised both by high industrial, urban and traffic emissions and continental climate. The examined data records consist of $O_3$, CO, NO and $NO_2$ hourly concentrations measured by the urban air quality monitoring stations. Local monitored temperature and forecast data are available from the meteorological office. Both classes of models have been identified on 1995-1998 and validated on 1999-2001 summer season data (from May to September). Grey box and neuro-fuzzy forecasts have been performed both on the maximum expected hourly $O_3$ concentration value one day in advance, and the

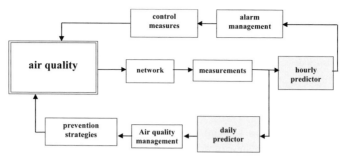

**Fig. 8.26.** The Decision Support System

maximum expected hourly concentration value for the afternoon at noon during the day itself. The capabilities of different predictors to foresee if the O₃ concentration will overcome an assigned threshold have been evaluated in terms of skill parameters, defined by the European Environment Agency. The indexes are estimated on the number of forecast and really recorded alarms:

- **SP** is the *percentage of correct forecast* smog events (probability of detection) (range from 0 to 100 with a best value of 100). The fraction of *unexpected* events is given by *(100 - SP)%*;
- **SR** is the *percentage of realised forecast* smog events (range from 0 to 100 with a best value of 100); the *percentage of false alarms* is given by *(100 - SR)%*;
- assuming an equal weight for smog events and non-smog event correct forecasting, the scoring parameters *SP* and *SR* can be combined to a *success index*, **SI**, ranging from -100 to 100 with a best value of 100.

The performance indexes, related to the O₃ threshold value of 140 μg/m³, have been estimated for both forecast models. The *persistent model* (tomorrow equal today) skill parameters have also been computed as lower bound performance indexes.

Fig. 8.27 compares the skill parameters computed for the four models: the daily (*grey box*) and hourly (*neuro-fuzzy*) predictors, their combination in the air quality system and the persistent model. The first two predictors have good performance in avoiding false alarms, while the *SP* index for the daily model claims for a forecast improvement in enhancing some episodes. The *neuro-fuzzy* predictor matches the persistent model in correctly forecasting smog events.

The air quality system, implementing the second internal feedback, improves operational effectiveness in the short term (hours), taking into account all recent available meteo-chemical measurements. The results underline the system synergy mainly in forecasting O₃ threshold exceedances (*SP*) and in global performing (*SI*). (Finzi et al.2000; Finzi and Volta 2000).

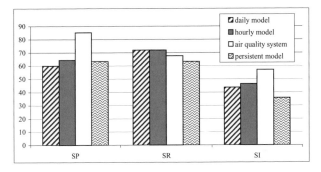

**Fig. 8.27.** The estimated forecast skill parameters

## 8.8 Long Term Modelling Simulation for Ozone Exposure Assessment

In order to assess the comprehensive effects of photochemical pollution, not only $O_3$ peak concentrations need to be examined, but also $O_3$ exposures on "seasonal" scale need to be quantified. Recent works (e.g. Kasibhatla and Chameides 2000) point out the importance to perform policies analysis on a "climatological" basis rather than focusing on a single critical episode; this approach allows both to evaluate better model performances (Hogrefe et al. 2001) and to quantify policies effects with respect to long-term air quality standards. Moreover measured concentrations available at given monitoring sites are generally not enough detailed to describe the spatial distribution of $O_3$ over wide areas, whereas this information is a crucial factor to evaluate the impact of photochemical pollution on natural ecosystems. Modelling systems can represent suitable tools for these purposes, allowing both the study of photochemical pollution with an adequate spatial detail and the assessment of appropriate emissions reduction strategies.

In order to estimate the $O_3$ concentrations long-term impact, cumulative approaches may be applied; these methodologies, for example, evaluate the number of exceedances over a threshold, defined with respect to the objective, both for health and natural ecosystems protection. At present time there is scientific concern about the evaluation of critical levels for $O_3$ effects on vegetation by means of the AOT40 index, which represents the cumulative exposure exceeding 40 ppb for daylight hours during a growing season. As for the health protection, following the WHO guidelines, the main concern is with the 8-hourly mean concentration, for which a threshold value of 60 ppb has been assumed. Some Authors (EMEP 1999) suggest to approximate the long-term evaluation of the 8-hourly average with the AOT60 index as surrogate indicator.

Referring to North Italy as case study, an integrated modelling system has been designed and implemented, including 3D meteorological pre-processor CALMET, a flexible emission inventory module POEM and two photochemical transport models, CALGRID and STEM-FCM. Initial and boundary conditions are obtained by means of a nesting procedure from the EMEP Lagrangian photo-oxidant model. Air quality simulations have been performed for the summer season 1996; in particular, STEM has covered the period from 1/5 to 31/7, while CALGRID has been run from 1/4 to 30/9.

In order to compare simulation results and monitored critical levels, some reference stations have been selected from the Lombardia Region network by using a statistical cluster analysis technique. The long-term run results have been compared with air quality data, in order to assess the model cleverness in actually rebuilding time and spatial features of pollutant concentrations. Simulation results over the summertime have pointed out a quite good agreement between computed and measured $O_3$ concentrations. Both modelling systems are able to replicate the overall behaviour of the measured $O_3$ concentrations, with a quite similar performance. In particular, the analysis carried out in terms of daily maximum values shows the capability of the modelling systems to actually reconstruct the different peculiar features of $O_3$ concentrations, so allowing estimates of realistic AOT40 and AOT60 indexes.

Fig. 8.28 shows simulated and observed AOT40 for crops evaluated at 8 monitoring stations. In case of missing measured data (in most cases less than 5%) AOT indexes have been adjusted by means of a simple fitting method. For both models the simulated indexes denote a general underestimation with respect to the observed values at urban sites as well as an overestimation at rural stations such as Gambara (with the exception of Varenna). Underestimation at urban sites is greater for STEM than CALGRID; this could be due to a slightly different parameterisation of vertical dispersion in urban areas, inducing a more intense $O_3$ consumption during evening hours. At remote stations STEM reconstructs AOT40 better than CALGRID.

AOT40 crop simulated fields are reported in Fig. 8.29. Both models exhibit higher values outside urban areas, both in northern and southern parts of the domain. Highest values occur in the portion of Pre-Alps near urban areas, and, in southern parts, near the Apennines mountains. Fig. 8.29 also reports AOT40 estimated by the EMEP Eulerian model (EMEP 1999), performing simulations over all Europe with a 50 km grid resolution. AOT patterns are quite similar, with the highest values placed in the same areas. Thanks to a greater horizontal resolution, STEM and CALGRID are able to better reproduce AOT40 variations, particularly over complex terrain areas.

Fig. 8.30 illustrates AOT40 and AOT60 levels for the 6-month period estimated with the CALGRID modelling system. The model results show most parts of the domain lying above the critical level threshold, both for forest and human health.

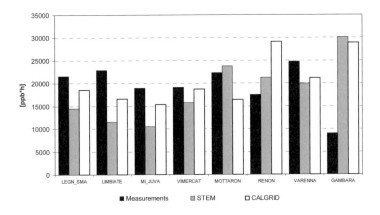

■ Measurements     ▨ STEM     □ CALGRID

**Fig. 8.28.** AOT40 for the period May-July 1996

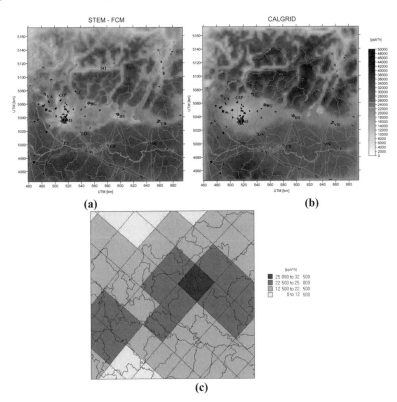

**Fig. 8.29.** AOT40 (May to July) fields: **(a)** STEM; **(b)** CALGRID; **(c)** EMEP (50x50 km)

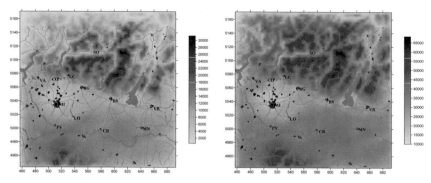

**Fig. 8.30.** AOT40 (left) and AOT60 (right) for 6-months period estimated by CALGRID

## 8.9 Conclusions

SATURN, via the activity of FOSEC (Formation of Ozone in South European Cities), has contributed towards a better understanding of photochemical pollution in South Europe. The related topics that have been treated are: the meteorological conditions that favour the formation of $O_3$ in specific South European areas, the effects of phenomena associated with different scale interactions on $O_3$ concentration in southern European cities, the emission of $O_3$ precursors, the indicators used for photochemical pollution and the associated sensitivity regimes, and finally the role of aerosols on the issue. For each of these aspects different questions were raised and answered through results reported by different groups within SATURN as well as other European projects (e.g. LOOP).

Further investigation with emphasis on ambient measurements is necessary in order to reduce the uncertainties associated with VOC-$NO_x$ predictions. Evaluation of emission inventories, $O_3$ production efficiency and removal rates of species other than just $O_3$ are required for the improvement of the numerical models. In addition, these should be combined with observation-based techniques in both urban and rural sites. On the other hand, predictions of the impact of reduced $NO_x$ and VOC on $O_3$ derived by 3D Eulerian photochemical models have large uncertainties. These uncertainties are associated with the difficulties in assessing accurately the VOC and $NO_x$ chemically sensitive regimes as well as in uncertainties in emission rates and meteorology. Further investigation is required for identifying the role of aerosols in $O_3$ formation.

# Chapter 9: Integrated Urban Air Quality Assessment

D. van den Hout[1], S. Larssen[2]

[1] TNO/MEP, P.O. Box 342, NL-7300 AH Apeldoorn, The Netherlands

[2] Norwegian Institute for Air Research, P.O. Box 100, N-2007 Kjeller, Norway

## 9.1 Introduction

Scientific development is very important for the progress of society, but not all science necessarily contributes to this equally. For environmental science it is probably safe to say that all types of research have, at least indirectly, application as their eventual aim. Application can be read here as contribution to the solution of environmental problems. More specifically, the results of environmental science can be expected to play - directly of indirectly - a role in decisions by environmental authorities and other stakeholders, including the general public.

Some scientific work directly applies to this societal process, but most work contributes indirectly, by adding to the complexities of knowledge, tools and data that as a whole makes practical applications possible. Fundamental research has explicitly no direct application aim; yet it provides the most essential building stones of the total edifice of scientific knowledge.

Many scientific projects have not been explicitly defined from the perspective of their place within an integrated structure. Ideally, the needs of research are derived from the application needs, by defining the problem first, then deriving questions to be answered, then defining the lacking information and finally setting up the projects which have to yield the lacking information. However, when following such procedures the difficulties in matching the demand for information and the supply of scientific expertise often results in scientific programmes that only partially answer the questions asked. Within SATURN a Framework Project was established to consider how the various outputs from this wide collaboration could be integrated to maximise the potential for assessing and managing urban air quality effectively.

### 9.1.1 SATURN's orientation towards integrated air quality assessment

SATURN has been a main platform of individual research efforts in Europe with urban air quality as its common denominator. It has similar characteristics as described above: it is a composite of contributions that together contribute to the advancement of urban air quality science, thus improving the competence of society to tackle air pollution problems.

The objectives of EUROTRAC-2 have not been restricted to the scientific realm, but practical application is also an essential goal. Reflecting this, the structure of SATURN has not only been defined from a scientific viewpoint: two important elements of SATURN structure feature its orientation to application (see Fig. 9.1). As the first element, the Main Group of activities (later redefined as cluster) 'Integration' was defined, which clusters the activities that focus on the application to air pollution management. The term 'integration' refers to the scientific activity of collecting and synthesising the scientific elements needed for shaping a direct interface with decision makers and society. The second element was the definition of SATURN's 'Framework Project', which aimed at linking SATURN to urban authorities and establishing interaction between air pollution policy makers and air pollution science. In order to have this as a focused activity,

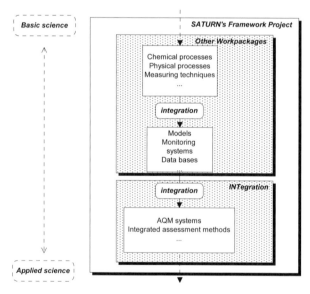

**Fig. 9.1.** Integration and application in SATURN. The INT(egration) Workpackage is based on the other workpackages (originally EXPerimental, MODelling and VALidation, later restructured as the Local Cluster and the Urban Cluster), which comprise scientific process studies and data bases and models based on those studies, which in turn form the basis for integrated assessment methods and Air Quality Management Systems. SATURN's Framework Project encompasses all contributions.

this SATURN level activity was defined as a project, to be funded on the basis of a well-defined programme. For a substantial part of its duration, funding was received through a grant from the Shell Sustainable Energy Programme. To investigate how the contributions of SATURN fitted together, the application chain given in Fig. 9.2 was designed, based on the structure in Fig. 9.1.

The structure shown in Fig. 9.2 was used to make an inventory of the aims of the various contributions in the first phase of SATURN in terms of its application orientation. Table 9.1 summarises the results. Not surprisingly, many contributions aimed at more than one element. All aims were addressed, with a tendency towards the more basic elements. A large majority of the contributions have improving insight/modelling/techniques concerning the air pollution processes (elements 1 and 2) among their direct aims. When for each contribution the direct aim having the highest number (the "farthest aim") is taken, the distribution of Fig. 9.3 is obtained, which indicates the directness of the application of the contributions.

Further analysis revealed that, at least in its first years, SATURN's contributions together could not be regarded as a fully linked application chain: the various contributions aimed more at applications beyond SATURN than 'downstream' in the chain within SATURN. This was due to the way in which SATURN (and EUROTRAC as a whole) was set up: SATURN operated rather as a platform of independent contributions than a set of linked contributions. It should be emphasised, though, that this applies to the planned applications only, not to scientific exchange. Inevitably, as SATURN activities developed, links emerged amongst

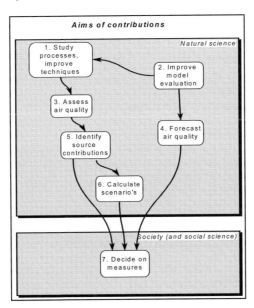

**Fig. 9.2.** Application chain of SATURN. The arrows indicate how the output of research in one field is needed as input for other fields. The total chain leads to direct input to decision making on air pollution.

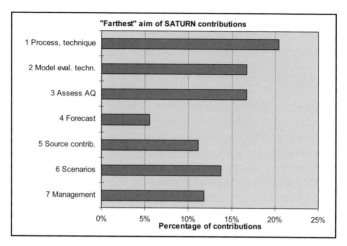

**Fig. 9.3.** Distribution of the aims per SATURN contribution that is farthest downstream in the application chain

several research teams as they forged close working relationships and implemented programmes with common aims and objectives.

**Table 9.1.** Application chain: possible objectives of contributions to SATURN

| *Objective* | *Addressed by % of contributions* |
|---|---|
| 1. To study processes and improve measuring and modelling techniques | 56% |
| 2. To improve techniques to evaluate models | 33% |
| 3. To assess existing air quality | 46% |
| 4. To forecast air quality for the next day(s) | 9% |
| 5. To identify source contributions to air quality by measuring or modelling | 25% |
| 6. To carry out air quality scenario analysis („what if") | 19% |
| 7. To decide and implement reduction measures | 11% |

## 9.2 Purposes and scopes of integration

### 9.2.1 Users' demands

Often the supply and demand of scientific results are not fully matched. Many scientists regard the advancement of science as their first mission, trusting that eventually their results will in some way be useful to society. Criteria for funding scientific research are often rather aiming at scientific excellence than application possibilities.

From the viewpoint of city authorities, however, application, and consequently often integration, is a must. As illustrated in Fig. 9.4, there are often numerous data and other forms of information, sometimes also models, available in cities, but these need to be combined and processed, usually by the municipal environmental department, before they can be used to give the information needed by policy makers, stakeholders and citizens.

In order to gain insight in the possible gaps between the supply by science and the demands of the users and to discuss possibilities of reducing the gaps, SATURN invited Dr. Frank Price from the municipal department of Sheffield (UK), an internationally well-recognised expert on urban air quality management, to present and discuss views on this. Among other things, he presented the principal conclusions of the *European Air Quality Management Project* (Daly 1998), and compared these with associated SATURN objectives (see Table 9.2). He presented analysis of gaps in the SATURN programme, again from the viewpoint of a municipal practitioner in air quality management (Table 9.3).

**Table 9.2.** Comparison of priorities for municipalities and for SATURN

| Key priorities for municipalities | SATURN Priority |
|---|---|
| Rules of thumb/normative calculations | Model development |
| Emission inventories/source apportionment | Model evaluation |
| What-if scenario development | Simulation |
| Linkage to policy/intervention development | Integration |

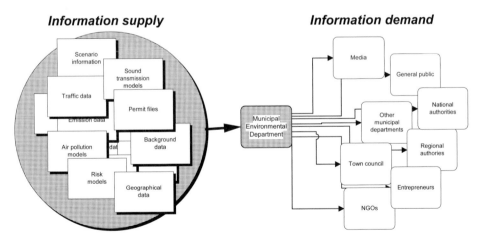

**Fig. 9.4.** Illustration of the information supplies and demands

**Table 9.3.** Gaps in the SATURN programme from the perspective of municipal authorities

| Actions needed at municipal level | Gaps in SATURN Programme |
|---|---|
| Wider dissemination of outputs | Communication between professions |
| User forums at regional level | End user involvement in project design |
| Mechanisms to study impacts and effect change | Integration of science with policy development |
| What-if modelling/scenario testing | Effects based approach to intervention strategies |
| Linkage of air quality to health | Personal exposure modelling |
| Linkage of air quality and environmental harm | Environmental critical loads |
| Tools accessible to lay people | Locality level monitoring /modelling |
| Confluence of ideas and action | Integration of science with locality management |

**Table 9.4.** Possibilities for improving integration

| Key issue for future air quality management | Possible central theme for action |
|---|---|
| There are real air quality problems in many urban areas | Best science needed to understand and promote change |
| Closer working needed across professional divides | End users / researchers to collaborate more |
| Full participation of those affected required | Science should come out of the laboratory and people must be brought into decisions making |
| Joined up action by all sectors | Participative working at all levels |

Table 9.4 gives possibilities for further integration of the SATURN programme and the active work of municipalities in air quality management. In the discussions it became obvious that there were important demands that could only be partially fulfilled by SATURN, as they often extend beyond the field of air pollution science.

## 9.3 Integration of elements

One might naively feel that integration of elements is not a scientific activity, but an engineering business. To some extent this is indeed true, as on the one hand basic, process-oriented research is by definition not included in this, and on the other hand the solving of operational problems often constitutes a major part of the work (the latter being a characteristic for almost all research). However, solving the practical problems requires good scientific insight in the various elements that are dealt with, flexibility and creativity. In the first place, the individual elements are often not compatible, in particular the output of one element is not directly useful as input to the next element, and solutions need to be found. Often, this requires conceptual changes in the elements. Secondly, when combining many elements into a larger system, the complexity may increase so much that simplification of elements is needed. 'Summarising' a complex element by simplifying it in the most effective way, when properly done, often requires deep insight in the essence of the processes within the element, sometimes even more that needed for development of the element itself. Thirdly, it is often very difficult, even for scientists within the same discipline, to fully understand the paradigms of the developers of the different elements – a very relevant example here is the decades old 'gap' that has been existing between measurers and modellers.

In the field addressed by SATURN, many combinations of elements are relevant. Several types of integration can be distinguished:

– Integration of elements within the scientific field of SATURN, *e.g.* measuring and modelling
– Integration of elements of SATURN with other air pollution research, *e.g.* urban and regional scale models
– Integration of SATURN's air pollution research with adjacent fields in the cause-effect chain: *e.g.* traffic modelling (causal side), exposure and health effects assessment (effect side)
– Integration with other fields that need to be taken into account in air quality management, *e.g.* costs of measures, synergy with noise management.

Below, several combinations are discussed in more detail.

*Mathematical models*

Models, being composite mathematical descriptions of coupled processes, are as such, powerful integration tools. Very specialised models may come close to the ideal of fully describing all knowledge of the process simulated, but more often models, even the very advanced ones, contain numerous simplified descriptions and assumptions. Within SATURN, modelling has been a major research topic. Chapter 6 serves as a comprehensive description of this topic within SATURN and hence is not discussed here further.

*Combinations of models and measurements*

Measuring and modelling often co-exist as alternative tools for assessing air pollution. However, the distinction between measurement and modelling is not as absolute as is often thought. There is no fundamental difference between assessment using measurements interpreted by common sense and assessment using models. Models may be described as mathematical formulations of one's understanding. Fig. 9.5 illustrates that there is an almost continuous spectrum of combinations of measurements and other assessment methods. Neither of the two extremes (a) and (h) is useful for assessing air quality: 100% measuring (i.e. doing measurements that are not generalised at all) gives incomplete information, while, at the other extreme, 100% modelling (i.e. applying models that have not in any sense been validated) gives unreliable information. So, a useful assessment comprises elements of both.

Measurement in combination with interpretation (b) has for many years been the most common way of assessing air quality. Measurements give point-wise information, and, therefore, interpretation of this information can only be conducted by taking into account various factors including the representativeness of monitoring sites.

A step further is the mathematical generalisation of measurements to give territory-covering information, such as spatial interpolation (c). This technique is useful for uniform areas with smooth air quality gradients between stations, but small-scale variations between stations cannot be identified. The interpolation can be improved by using relationships between the air pollution levels and geographical characteristics. Furthermore, for selected key-parameters (regarding source magnitude, meteorological conditions, configuration) empirical relations with air quality levels can be established. This technique uses the key-parameters for the interpolation instead of the physical distance in case of spatial interpolation. When the relationships between air quality levels and local characteristics have a fair amount of detail, they can be regarded together as constituting an empirical model (d).

A promising technique is data assimilation (e), which adjusts the values of the uncertain model parameters and measured results to find the best fit of the concentrations to the entire set of measurement and model data. The resulting map of concentration matches neither the original measurements nor the original model results, but it gives the mathematically best approximation of both.

Finally there is the sole use of validated models (f and g). Not only the reliability of the dispersion model, but also the quality of the emissions and dispersion input parameters need to be taken into account.

Within SATURN, a great deal of attention has been given to the techniques centring around measuring techniques (interpretation, interpolation), see Chapters

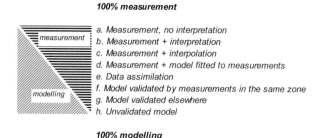

**Fig. 9.5.** Combinations of measuring and modelling

4 and 5, and to the techniques centering around modelling (validation, which has, for a large part of SATURN's existence, even been a Main Group of activities), see Chapters 6 and 7. However, there have been no contributions in SATURN in which an equally balanced combination of measuring and modelling techniques such as data assimilation was developed. This is remarkable, as the new European air quality directives encourage air quality assessment using a combination of measurements and models, and several EU member states are now developing such methods. As these methods seem to be often in an early stage of development, it is to be expected that these techniques will be extended. Especially for describing the complex air pollution fields in cities these techniques will be very valuable, but at the same time the complexity of these urban patterns poses major computational challenges.

*Integration of spatial scales*

It is very significant that SATURN has been divided into a local scale and an urban scale cluster, and that SATURN has been defined as the urban project of EUROTRAC. In air quality research, there are major differences between the local, urban, regional, continental and global scale. On the other hand, problems of high pollution levels are often due to a combination of contributions from these scales (see also Chapter 2): a local contribution by a near source plus the contributions of sources in the same city plus the contribution from more distant sources in region, continent and the rest of the world. In SATURN, much attention has been given to stimulate the interaction between the groups working at the local and the urban scale respectively. Also, from the outset of SATURN, it was realised that regular interaction with GLOREAM, the EUROTRAC subproject dealing with the regional scale, was very important. In spite of this, however, there are still major problems in integration of models on these various scales. Straightforwardly linking advanced models of different scales, which may on their own already approach the capacity limits of today's computers, could result in unacceptable computing times. In the EU's CAFE programme and the Convention of Long Range Transboundary Air Pollution the inclusion of urban air quality in the integrated assessment modelling, which until now has only dealt with the regional and continental scale, is regarded as a very difficult task. In SATURN, solutions have been found by either simplifying the models or their input, by decoupling the models, keeping them mutually inconsistent to some extent or by modelling only representative samples of the smaller scale domain.

*Integration of models for research and models for policy support*

Although the combination of modelling for research and for policy support is not a very obvious combination of elements, it is worth mentioning here, as it poses substantial practical difficulties. Many models, particularly the most advanced ones, aim at describing the processes that are simulated in the best possible way. When these models are applied to e.g. episodes covered in a measuring campaign, they can be very important for evaluating the status of the scientific knowledge. When these models are applied for policy-making purposes, the problem is often that the models are too specialised, too limited in scope to be useful. For policy support, a model should be able to calculate the parameters that are important for the policy makers, viz. the parameters of air quality limit values, usually expressed in annual statistics and applicable with a high spatial resolution. This is particularly pressing for developers of Air Quality Management Systems (AQMSs), where sometimes compromises regarding spatial resolution, temporal coverage and number of sources are needed before a model can be made operational in such systems (see Section 9.4). But also for stand-alone models such simplifications are often needed. A good example is the OFIS model (see Chapter 6).

*Models and databases*

A very common issue is the integration of models and their input. In suites of models, part of the input is formed by the output of the 'upstream' model, but even there also input data are needed. Standardisation of input data (on emissions, air quality, meteorology and topography) is very important and has, for emission and air quality data, been addressed throughout SATURN (see Chapters 1 and 3).

*Integration of ambient concentrations with exposure of the population*

The exposure of sensitive targets is a better, but more difficult, proxy of air pollution risks than ambient concentrations. Within SATURN, some contributions dealt with human exposure.

The AirQUIS AQMS system contains a module for computing population exposure distributions based upon home addresses (see Section 9.6.2).

Kukkonen et al. developed the model EXPAND for the evaluation of population exposure, processing and visualising the computed results using the Geographical Information System (GIS) MapInfo. The population exposure model consists of two parts: exposure in traffic and exposure elsewhere (work, home, other). The exposure in traffic is computed separately for each street section, and the exposure elsewhere is computed in a grid with spatial resolution of 100 m. The population exposure model has been applied in the modelling of population exposure to $NO_2$ in the Helsinki Metropolitan Area (see references in Section 9.6.8).

Colvile has collected one of the largest available data sets of exposure of road-users to air pollution in Central London, along three fixed routes for bicycle, bus, and car users, and along random routes for cyclists. This confirmed that road users are exposed to levels of air pollution significantly higher even than roadside measurements. Route and weather conditions are strong determinants of exposure. Mode of transport is a much less significant determinant. Colvile also made initial attempts to calculate road user exposure, carried out using the operational air quality modelling system ADMS-Urban. The model was able to reproduce the variability in exposure as a function of route and weather conditions, but had difficulty reproducing the observed enhancement of road user exposure over roadside concentration. This was probably due to the lack of high spatial and temporal resolution in the emissions and receptor position data. The potential importance of street canyon intersections in determining exposure, due to the high emissions density and extended periods of time spent there in congested traffic, was highlighted by this modelling exercise. This indicated potential to reduce the impact of traffic emissions on human health more effectively than by emissions reduction alone, by combining emissions reduction with traffic management measures that alter position on the road and time spent at the most polluted locations.

By the end of SATURN, Colvile worked on integrating the street canyon intersection dispersion modelling work with exposure assessment including microsimulation of traffic flow. The potential importance of highly localised short duration peak concentrations requires time-dependent large-eddy simulations.

*Integrating air quality models with cost modules*

Obviously, decision makers need to know the costs of measures. For some types of measures, cost indications can be given and they can be linked to reduction options in the database of an AQMS. Other measures are so intricately linked with other policy fields that it is impossible to isolate the cost of an air quality measure from all costs. Work on integration of cost modules into AQM systems has only started. Tthe AirQUIS system has been used to develop abatement strategies based upon cost-effectiveness in reducing the total population exposure.

*Integrating air quality models with models for polluting activities*

Integrated systems often have scenario modules that are able to generate emission scenarios, based upon activity and consumption data. These can for instance be a set of multiplication factors to be applied to emission factors, may be only for a given area in the city, which quantify the change in emission over the years. In SATURN, some contributions have gone further than this, by including a model that calculates the development of the polluting activities i.e. traffic distribution models (cf. Section 9.4).

*Integrating air quality modelling with models for other environmental fields*

For city situations, decisions on measures to control air pollution often have substantial impacts on other environmental fields, in particular noise and, less often, external safety. Also the data needed for assessing levels of air quality, noise and safety risk have similarities – many are even identical. Within one contribution to SATURN, the fully integrated AQMS URBIS was developed for air quality and noise, and at a later stage external safety was partially included as well (see Section 9.6.9). This system features the high resolution assessment needed for calculating compliance with limit values for air quality and noise. Further integration, e.g. with the fields of spatial planning or transport planning, has not been part of SATURN's activities.

*Integrating air quality assessment with public information*

The political motivation for measures is not only determined by legislative restrictions, but also the public awareness can be an important reason to put air quality higher on the political agenda. As air pollution can usually not be seen or heard, and often not smelled, public information is more important than for e.g. noise. Scientific reports on air quality can, however, be difficult to comprehend by the general public. Because of this, it is important to present, where useful, the results of air quality assessment in a way that is digestible for the public. In SATURN, most of the attention for this has been given in relation to the development of AQMSs. A related development took place in the APNEE project (funded by the EU IST programme), where SATURN groups participated (AUT, NILU, UPM Madrid) and which also included participation of mobile telecommunication companies such as France Telecom and Telefónica (Spain). In this project information on air quality was provided by using Internet and also the mobile telephone technology (URL 9.1). The citizen could subscribe to different services (e-mail, newsletter, SMS, etc.) by using an Internet form to receive accurate information on air pollution monitoring and forecasting data. In addition, a WAP interface was developed to provide the results of city air pollution monitoring data and forecasts. Such exercises were carried out in several cities (e.g. Madrid and in Norway).

## 9.4 Urban Air Quality Management Systems (UAQMS)

### 9.4.1 User needs to UAQM systems

For today's environmental authorities/ managers there is a strong need for systems that enable them to efficiently perform their main task: to secure, through planning and abatement decisions, a continued acceptable or improved air quality, or development towards compliance with directives, standards or guidelines. With the emphasis put on the efficiency of environmental control efforts, the "systems" referred to here are 'operative' software systems. Efficiency is enhanced, the more self-sufficient, user friendly and functional such UAQM systems are. There is a trade-off, however, between a high degree of system operability by non-scientific users (e.g. by city environmental control departments), and the degree of accuracy, comprehensiveness and relevance of results provided by the system. Research oriented tools operated by scientists in response to specific requests for analysis and assessments would in general provide results of higher quality. However, the discrepancy in quality between research tools and self-sustained systems is being diminished, as the software programmes and hardware platforms become more powerful. For many applications, self-sustained systems operated by people outside the scientific community give results of a sufficient quality.

There is a range of needs that state-of-the-science UAQMs should satisfy. Urban environmental authorities and managers of institutions with the operative responsibility for air quality management, to assess, control and improve air quality, need systems that can satisfy the following:

- Near-real-time access to monitoring data, and direct presentation of this to the public in an understandable format (now using Internet or other telecommunications modes);

- Short-term forecasting of air quality ("tomorrow's air"), and similarly presentation of this to the public. Also as a basis for activating short-term abatement actions;
- Assessment of present air quality (in a statistical sense, e.g. last year's air quality, its variation in time and space, exceedances of limit values, etc.), as needed for instance for reporting to higher authorities (e.g. the European Commission). Spatial and temporal resolution according to the nature of the problems;
- Planning needs: Prognosis (forecasting) of future air quality, for various scenarios of development and abatement;
- Development of cost-effective abatement strategies, where costs of abatement are compared with avoided damage costs (benefits). Possibilities for developing optimised abatement strategies;
- Visualisation tools to support the needs for presentation of results from the system, for the system users as well as for the public, and in various media.

Traditionally, urban air quality management work has been time- and resource-consuming, due to the large amounts of data needed for the analysis, as well as the equally large amounts of simulated results. In parallel with scientific developments in the various air quality related fields (dispersion models, knowledge of emissions and their determination, monitoring methods, data quality assessments etc.), software systems have been developed, which more or less efficiently link the various modules needed, so that analysis can be carried out in a less resource-consuming way. GIS tools have been integrated in the systems, facilitating presentation and visualisation of results geographically. Internet and other modes to make data public has been utilised as well.

Development of such software tools has been carried out in a multitude of scientific institutions and in the industry. This development has been driven by various forces, such as:

- Special needs related to area-specific pollution problems, and local authorities have developed cooperation with research groups in the area;
- A more general starting point where the air quality management needs of cities, as a result of requirements set by European/National legislation, is realised.
- Research groups develop their ideas based upon their expertise and experience.

Thus a multitude of different urban AQ management systems have been developed, which are in various stages of comprehensiveness, area-specificity and self-sufficiency (independence of use by others than the developers). The Integration cluster of the SATURN project has provided a forum for exchange of ideas, concepts and development presentations that has been fruitful for the participants. The exchange in this forum has shown that the various systems have advantages and shortcomings, and provided feedback which the development teams have used as guidance in their further development efforts.

No doubt, at present there are viable UAQM systems available to users in cities, either as self-sufficient systems for use by the cities themselves, or to be used by the research groups developing them on a contract basis in response to specific needs by cities. The scientific quality and operational functionality of the systems is high in general, and has already significantly improved the ratio between the extent of customised results and the resources used. However, the technological development related to software and telecommunications gives the possibility for continuously enhancing this ratio between effective results and resource input.

### 9.4.2 Structure of UAQM systems

The key feature of a modern environmental information and management system is the integrated approach that enables the user in an efficient way not only to access data quickly, but also to use the data directly in the assessment and in the planning of actions. The demands to the integrating features of the systems, to enable monitoring, forecasting and warning, and future strategy planning, as well as visualisation and presentations, will be increasing in the future. The typical structure of UAQMs responds to this demand for integration.

Taking THE USER as the starting point, the one who is delivering the terms/premises for the development and functionalities of the UAQM systems, Fig. 9.6 visualises how all the various typical elements (modules) of an air quality analysis (emissions, monitoring data, models etc.) serves him/her as they are linked together in software systems, the main purpose of which is to support the needs for making decisions, be it in the short term or long term.

There are a large number of elements and types of data and models that are to be integrated in a functional AQM system. Fig. 9.7 indicates the typical elements of data and models, and how they need to be linked through an interface, which includes a GIS tool, and also a report generator and visualisation modules (called "data wizard" in the figure), to provide the needed outputs. The data wizard module of course will have many general elements, but is also a place where the AQM system can be custom made to the user, to fulfil his/her special needs.

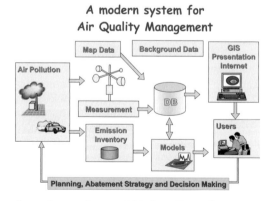

**Fig. 9.6.** Structure of a modern environmental information and management system

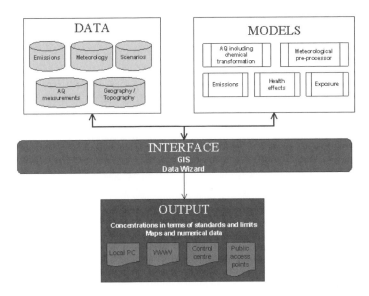

**Fig. 9.7.** Typical data and system elements and modules to be integrated in an AQM system

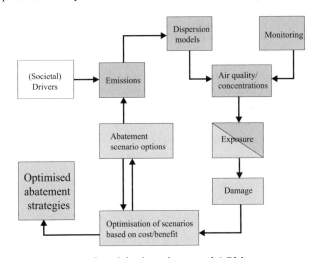

**Fig. 9.8.** Functional structure of modules in an integrated AQM system

Fig. 9.8 shows the typical structure of AQM systems in a more functional way. The various modules are linked (integrated) in loop structure. The "emissions" module is often taken as the starting point of an air quality analysis. However, the "monitoring" may be an equally valid starting point, to first assess present air quality before entering into its analysis. The point is that in the integrated loop type structure, the system can be entered from different angles.

There are (at least) two main sections of the loop:

- the *air quality assessment* section (dark grey boxes), where the air quality in an area (e.g. a city) is assessed either by monitoring, by modelling, or a combination. Most UAQM systems include these modules, and results to be visualised may include e.g. present concentrations (on-line or statistically, measured data in points or isolines of modelled concentrations), forecasts, baseline prognosis.

- the *air quality abatement* section (light grey boxes), where damage and its costs, and abatement options and their costs are assessed and compared /optimised. So far few UAQM systems include much of this section in an integrated way, although present – day systems are suitable for making the same kind of analysis "off-line", i.e. the AQ assessment part is activated and run for various abatement strategies, and results compared and "optimised" into control packages/strategies, external to the system. Collaboration within SATURN has resulted in a major project OSCAR (Optimised Expert System for Conducting Environmental Assessment of Urban Road Traffic) to be funded by the European Commission under the Framework 5 programme (Key Action: City of Tomorrow) which will lead to an air quality management system with an integrated capability to assess the effectiveness of various measures for improving air quality in and around streets.

### 9.4.3 Synthesis and summary of UAQM systems in SATURN: developments, applications and experiences

A number of research groups have participated in the Integration cluster of the SATURN project, which have had sustained development of AQM systems throughout the project period. Most of them came to the SATURN project with already partly developed systems. This development continued during SATURN, driven by the ideas, enthusiasm and funding that was within each group. It can be anticipated that the forum created by SATURN provided an additional impetus to development, and that there was cross-breeding of ideas to the benefit of the participants.

In Section 9.6, each of the participating systems is described briefly, as well as their development during the SATURN period, and the applications that each group put their systems to. The systems can be classified into approximate groups according to their level of self-sustainability and availability for outside users:

– *Research-tool systems:*

These are systems used by the research groups where they were developed, for research or applications in specific areas, on a project or contract basis. Some of these systems are intended to be self-sustained systems, when fully developed.

– *Partly self-sustained systems:*

These are systems which in given circumstances can be used or accessed by users, in cooperation with the developer group.

– *Self-sustained systems available on the market:*

These systems are available on the market, to be installed at platforms at the users place, and used by them after sufficient training. These systems are continually being further developed and improved.

A synthesis and summary description of the SATURN-participating UAQM systems is provided in the following. As the systems are complex, have a multitude of modules and submodules incorporated in them for specific uses, and applications are many, this summary touches only upon the really main features of the systems. The reader should refer to Section 9.6 and the references there, to obtain a better and fairer impression and knowledge of the various systems.

### About the systems

There have been eleven UAQM systems participating at SATURN. In the following they are listed alphabetically according to name of system:

- ADMS-Urban:       *ADMS-Urban Air Quality management system* (Cambridge Environmental Research Consultants)
- AirQUIS:          *Air Quality Information and Management System* (Norwegian Institute for Air Research)
- GAMES/AQUAS:      *Comprehensive Modelling and Decision Support Systems for Photochemical Pollution Control in Metropolitan Areas* (Univ. of Brescia, The Electronic for Automation Department Group)
- IUAQMS:           *Integrated Urban AQM System* (Univ. of Aveiro, Dpt. Of Environment and Planning)
- IUEMIS:           *Integrated Urban Environmental and Information Systems* (Aristotle University of Thessaloniki, LHTEE Group)
- OPANA:            *Operational Atmospheric Numerical Pollution Model for urban and Regional Areas* (Technical Univ. of Madrid, Environmental Software and Modelling Group).
- OPUS-AIR:         *Integrated Assessment policy system for European Urban Air Pollution Studies* (Aristotle University of Thessaloniki, LHTEE Group)
- Photosmog:        *Photosmog pollution Episode Warning System* (UFZ-Centre for Environmental Research Leipzig-Halle Ltd, Dpt. of Human Exposure Research and Epidemiology)
- UDM:              *Urban Dispersion and Exposure Modelling System* (Finnish Meteorological Institute)
- URBIS:            *Urban Information and Management System* (TNO)

Several other AQM systems exist, in Europe as well as in the USA and probable in other parts of the world, which have not been participants of the SATURN subproject. The groups active in the Integration cluster know of many of these, but it has not been the intention here to make a full overview of all existing systems.

The participating systems can be classified as follows:

– *Research-tool systems*: GAMES/AQUAS, OPUS-AIR, Photosmog, URBIS, UDM.
– *Partly self-sustained systems*: IUAQMS, IUEMIS, OPANA.
– *Self-sustained systems*: ADMS-Urban, AirQUIS.

A synthesis of the capabilities of the systems:

– Almost all of the participating systems have the capability to produce assessments of air pollution and its variation in space and time, as well as analysis of effects of abatement strategies. These results are produced applying meteorological and dispersion/chemical transformation models of various types and resolution.
– Some of the systems have modules enabling air pollution monitoring data to be displayed on-line, some also on the Internet (e.g. AQMS-Urban, AirQUIS, IUAQMS).
– Many of the systems also operate on-line to give short-term forecasts of air pollution (1-2-day forecasts).
– The systems are more or less integrated, in terms of operative coupling of various modules (e.g. emissions/meteorological/dispersion/transformation/exposure modules).
– Most systems (but not all) operate on PC platforms.
– Some systems have been developed far enough that they can be used by trained outside users (e.g. urban/local AQ officers) (e.g. ADMS, AirQUIS).
– Transport/traffic models (specific ones, not in general) are incorporated in some systems
  (e.g. IUAQMS, UDM).
– Meteorological flow field models (e.g. CALMET, HIRLAM, MM5, MEMO) are incorporated in most systems.
– Photochemistry modules are included (in e.g. GAMES, OPUS-AIR).
– Submodels for street-level pollution are incorporated in many systems (e.g. ADMS-Urban, AirQUIS, IUAQMS, URBIS, UDM, OPUS-AIR).
– Population exposure calculation modules are included in some (AirQUIS, UDM, URBIS).
– Noise calculations are included in one system (URBIS).
– Internet applications related to the AQM systems (access to the systems via Internet, as well as input of data and visualisation and presentation of results, pollution forecast information and warning, etc) have been developed by some (e.g. OPANA, IUEMIS, AirQUIS), and this development is progressing rapidly to more advanced applications.

- Development of broad Internet portals for environmental information (e.g. IUEMIS).


## Developments under SATURN

Although most of the systems were at a more or less advanced development stage at the start of SATURN, a number were further developed during the SATURN period. The efforts of the various participating research groups during SATURN resulted in system modifications and improvements related to:

- integration (linking) of modules;
- flow/chemistry/dispersion/exposure models;
- user friendliness / operability;
- system capacity / speed (related also to the general improvements in hardware specifications);
- evaluation/validation of systems and modules (using data bases of measured data from experimental campaigns;
- extensions to more pollution compounds;
- telecommunications / Internet applications.


## Applications of the systems

This is described in further detail under each system in Section 9.6. The general picture is that all of the systems have been actively applied in studies corresponding to one of the main purposes of UAQM systems: evaluation of effects of abatement scenarios. Examples of such applications:

- ADMS in a range of cities in the UK, in part operated by urban users. Also in cities in China.
- AirQUIS in four cities in Norway (forecasting and abatement scenarios). Also in cities in China.
- GAMES/AQUAS in the Greater Milan area.
- IUAQMS in Lisbon (together with similar assessments in Gdansk, Geneva, Genoa, Thessaloniki, Tel-Aviv, as part of the SUTRA 5th FP project).
- OPANA in Madrid and other cities in Spain, as well as Leicester, and in Equador.
- OPUS-AIR in Greater Athens, Milan and Stuttgart.
- URBIS in cities in the Netherlands.
- UDM in Helsinki and other Finnish cities.

*Summary*

It is fair to state that the work and activities carried out in the Integration cluster has resulted in further development of air quality management systems, and in gathering of essential experience through the applications carried out. This experience also includes experience by outside users of the self-sustained systems, e.g. by city authorities.

Integrated Urban AQM systems are now routinely used by urban authorities and other users as they provide capabilities to display data on-line, perform assessments with time and spatial resolution of present and future air pollution distribution attribute the sources contributing to the air quality and to test and formulate effective abatement strategies. The use of the systems by outside users does require a large amount of training. Different systems have different capabilities, so an evaluation of user needs is necessary when acquiring of a system is contemplated. As mentioned earlier, other systems than those described here are available.

The development of AQM systems will continue. Models will improve, systems will be extended to more compounds (e.g. various PM size fractions), operability and user-friendliness will improve. This development will be driven partly by the further demands to AQ management from national and European authorities, but also by the science itself, and its impetus towards improving the usefulness of its products.

# 9.5 Outlook

SATURN has included a wide variety of contributions, ranging from basic research – aiming at better understanding the processes causing air pollution – to integrative research – often aiming at application for policy support. A substantial number of contributions had integration as their primary focus. These contributions were developed largely in parallel, but SATURN has provided an important platform for exchanging insight and experience. Within SATURN, several initiatives have taken place from which collaborative EU projects have emerged: APNEE, ISHTAR, OSCAR, ENV-e-CITY and SAPPHIRE. This momentum in new urban air quality research has manifested itself in the Cluster of European Air Quality Research (CLEAR) consisting of several projects funded as part of the Fifth Framework Programme in the Key Action City of Tomorrow.

The integrative work centred around AQMSs and included extension of these systems with new data and models, analysis of the uncertainties in the results, increasing the efficiency of the systems, improving the user-friendliness, adapting the systems better to the needs of users. The activities were not limited to improvement of given systems, but there were also more basic activities such as rethinking the concepts of AQMSs and reconsidering their purposes. On the one hand a convergence of the approaches and techniques could be seen, which was caused by the gradually stronger European component in the field (both scientifically due to the SATURN collaboration and politically due to the increasing impact of EU legislation). The systems tended to address more the air quality in terms of the European air quality directives. The databases, and to some extent modelling techniques as well, became more harmonised. On the other hand, the new ITC techniques that became available provided possibilities that are still in the exploration phase. The AQMSs were connecting also to fields adjacent to the core field of SATURN: exposure of population and related environmental fields. Hardly any progress in the addition of methods to assess costs of measures could be noted, however.

It is to be expected that further air pollution research will affect all fields within SATURN. Scientific insight into the basic processes will progress, and this basic insight will be gratefully included in more integrative approaches such as models and AQMSs. Consequently, the quality of integrated approaches and tools will improve, but also the integrative techniques themselves will become more advanced, not in the least due to the ever advancing IT technology.

There is always a gap in detail and scientific sophistication between the basic research techniques and the more integrated approaches, as it takes time to transfer knowledge from the basic level to the integrative level and, furthermore, simplifications are often needed to keep integrated systems manageable. However, IT technology will probably continue its rapid evolution, benefiting the integrative approaches probably even more than the basic research techniques, and so there is reason to believe that the gap will decrease. The rapidly developing ITC technology also improves rapidly the ability to provide the public with on-line updated air pollution information.

The incentives for integrated assessment will continue to exist, and may even grow stronger. At the European level, there is now a tendency of leaving single-issue policy making in favour of more holistic approaches. A major motivation for this is that the obvious and cheap measures have now been taken, and that further measures to reduce air pollution problems will be complex (and perhaps expensive) and will usually affect more than air pollution alone. This notion has lead to the establishment of the Clean Air For Europe (CAFE) programme, in which integration of policy fields is a major objective. This change in attitude also brings along an increasing interest in costs and effects of air pollution. As more advanced and subtle approaches are becoming worthwhile for identifying the most effective way for society to reduce the harmful effects of air pollution further, the interest in extending air quality assessment with exposure assessment, risk assessment and cost effectiveness seems to be growing. Consequently, it is to be expected that models and AQMSs will be further extended with cost, exposure and risk modules.

It also seems probable that integrated assessment will, sooner or later, include more than air pollution alone. As stated earlier, many measures to improve urban air quality affect also other environmental themes, such as noise and external safety. The most notable example is road traffic, which is in many cities now the main source of air pollution. Particularly in cities, where generic air pollution strategies and legislation have been implemented in practice, it is important to investigate the full range of the impacts of measures. In addition, the current air quality legislation and the upcoming noise legislation have many similarities, particularly regarding the assessment of urban levels. Consequently, it is to be expected that AQMS will evolve towards more comprehensive Environmental Management Systems. There are also possibilities for further integration of environmental models with methods in the fields of transport and spatial planning. Traffic models can be (and have already been) easily extended with emission modules and GIS tools to make spatial manipulations in databases easier than it used to be.

As mentioned above, further work needs to be directed towards the integration of measurements and models, such as data assimilation. It is expected that the new EU air quality legislation will result in the development of fully operational techniques for the regular territory covering assessment of air pollution that is required under the new directives.

# Annex: Description of UAQM systems participating in SATURN

The systems are listed alphabetically, according to their name (acronym).

## ADMS-Urban: ADMS-Urban Air Quality Management System

*Developer*:Cambridge Environmental Research Consultants

*Provider*: Dr David Carruthers, enquiries@cerc.co.uk, www.cerc.co.uk

*Application areas:*

1. Urban air quality assessment and forecasting taking into account the full range of emission source types including: road traffic, industrial, commercial and domestic emissions.
2. ADMS-Urban is used to assess compliance with air quality objectives, for current and future scenarios, on behalf of local government authorities and national government.
3. It is also used for planning and policy support for the organisations including the UK Department of the Environment e.g. assessment of exceeedence of UK National Air Quality Strategy objectives, EU air quality standards and proposed EU standards.

*Structural elements included/other main features*:

1. Three-dimensional quasi-Gaussian model nested within a trajectory model. ADMS-Urban models the dispersion in the atmosphere of passive, buoyant or slightly dense, continuous releases from single or multiple sources which may be point area or line (road) sources.
2. The model uses an up to data parameterisation of the boundary layer structure based on the Monin-Obukhov length $L_{MO}$, and the boundary layer height h.
3. ADMS-Urban can be linked to EMIT (Emissions Inventory Toolkit) for calculating and manipulating emissions of toxic and greenhouses gases, which may be derived from EMIT's emission factor datasets or may be user-specified.
4. $NO_x$ chemistry modelling including grid-based trajectory model to account for the chemical reactions between the background data site and the domain of interest
5. Full integration with a Geographical Information System (ArcView and MapInfo) allowing easy emission set-up, output presentation and analysis.
6. Links to pollution concentration monitoring data for display of results against monitored data and meteorological forecasts for air quality forecasting
7. An integrated meteorological pre-processor which calculates the required boundary layer parameters from a variety of input data e.g. wind speed, day, time, cloud cover or surface heat flux and boundary layer height.
8. A range of modules allow for the effects of plume rise, complex terrain, street canyons, and buildings.

*Technical platform:*

High performance PC platform with the following minimum requirements:
(i) Pentium III processor
(ii) 256B RAM
(iii) 12GB hard disk
(iv) suitable monitor and accessories e.g. CD drive, colour printer

*Developments during SATURN:*

During the SATURN program ADMS-Urban has been developed in several ways:

- It is now more closely linked with EMIT emissions inventory toolkit allowing a two-stage rapid assessment of mitigation strategies. The user can assess the effect of traffic management strategies on overall emissions before then investigating the effect on air quality
- A more complex chemistry scheme including more reactions than the 8 reaction GRS (Generic Reaction Set) scheme has been developed and tested for implementation in the model
- The significance of a mid-scale dispersion length scale, that might be called "neighbourhood scale" and lies between city-scale and street-scale has been investigated by analysis of field experiment tracer results and urban morphology.
- The scope of the model has increased with an increased number of sources handled and more options for the user
- The ADMS-Urban and ADMS-Urban/EMIT UAQMSs have been used by an increasing number of groups during the SATURN project. They were used by

over 70 local authorities in the UK for the first round of Air Quality Review & Assessment.

*Applications of the ADMS-Urban system:*

1. **Planning:** The ADMS-Urban system is being used currently, on behalf of the UK government, to investigate current and predicted air quality in several of the UK's largest cities. Analysis of current and future air quality in London is underway, with investigation of the impact on air quality of proposed traffic management strategies a key area of investigation.
2. **Forecasting:** ADMS-Urban can be run as part of an integrated system which automatically downloads meteorological forecasts and monitoring data and generates pollution forecast maps which are posted on the internet. Calculations are carried out each day on receipt of the forecast met data. It is planned to extend the system to use hourly traffic counts. See www.cerc.co.uk/avon/ for an example of the air quality forecasts.
3. **Mitigation Strategies:** The ADMS-Urban/EMIT system is in use in 5 cities in the industrial province of Liaoning in northeast China. For the heavily industrialised city of Fushun, the model has been used to assess the impact of various mitigation strategies such as reduction in the percentage sulphur in coal, clearing of single-storey dwelling, introduction of district heating and industrial restructuring, on air quality. The focus is on industrial sources and heating sources rather than traffic. This is assessed both in terms of health impact and a full cost-benefit analysis, in order to inform the decision-making by the city and regional governments.

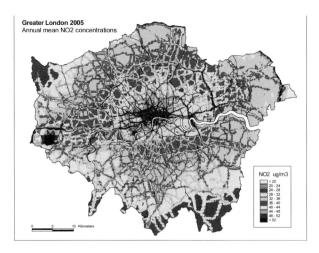

**Fig. 9.9.** ADMS-Urban detailed contour plot of modelled $NO_2$ annual average concentrations for London (1999)

## AirQUIS: Air Quality Information and Management System

*Developer*:  The Norwegian Institute for Air Research (NILU) group at Kjeller

*Provider* Contact/more information: Steinar Larssen / Trond Bohler.
http://www.nilu.no

*Application areas:*

The system is suitable for:
- assessments of present and future urban air quality and its spatial and temporal variations. This is enabled both by means of its monitoring data module which can provide near-real-time and statistical presentation of time series of pollution compounds and meteorological data, as well as by means of modelling of pollution concentration and population exposure fields;
- short-term forecasts of urban air pollution, when coupled to a meteorological forecast model;
- planning tool: development of abatement strategies. Basis for development of cost-effective abatement strategies.

*Structural elements included/other main features*:
- Modules for monitoring data, emissions inventory, meteorological/wind field models, urban scale dispersion models which includes point and line (traffic) source submodels for hot spot receptors, population exposure model
- The system is fully database-coupled GIS-based. The modules for emissions, wind and dispersion models and exposure models are fully integrated. Well-developed functionality for entering and modifying input data for the emissions calculations, their spatial and temporal distributions, and their aggregation (last item under development).
- Visualisation module enables time series, two-dimensional maps (isolines, values in grids/shape areas).
- Coupled to Internet applications, which enables near-real-time presentation of monitoring data, as well as modelling results. Remote operation.

*Technical platform:*

The system is operated from a PC server platform with the following minimum requirements:
- PC Pentium III
- 256 MB RAM
- 3 x 4 GB hard disks
- Suitable monitor and accessories (e.g. back-up units).

Requirements to PC clients capacities are less.

*Developments during SATURN:*

Development and improvements of the system is on-going, with emphasis on modularisation of its central data base, to increase operation and calculation speed and to facilitate a simpler process of further improvements of the system.

During SATURN emphasis has also been on user friendliness, as well as incorporation of items necessary for meeting requirements to AQ assessments which are given in the EU AQ Directives: e.g. calculation of directives-relevant statistical values/percentiles, QA/QC procedures, alarm functions when monitoring levels exceed alarm values.

Also, new telecommunications modes have been utilised in the data presentation module, to enhance the contact with the public. Push/pull telecommunications modes are used, such as mobile phone SMS and Internet for this purpose.

*Applications of the AirQUIS system: Examples:*

- The AirQUIS system is being used for AQ forecasting (24-48 hours, $NO_2$ and $PM_{10}$) in 4 cities in Norway (incl. Oslo and Bergen). The meteorological forecast is given by the MM5 model, and the system is run every day during the winter season, providing both present state of air quality relative to limit values and the forecast for the next 2 days.
- The system has been established in the Grenland industrial area in Norway as part of the APNEE research project (under EU 5th FP). The system gives preprogrammed SMS messages to selected subscribers, with essential information about the present air quality in the area.
- The system is being used in various fashions to determine the effect in urban areas (on concentrations and population exposure) of specific abatement measures and strategies, such as, e.g. the effect of studded tyres relative to that of small-scale wood burning for space heating, on the $PM_{2.5}$ and $PM_{10}$ concentrations in Oslo; and the development of cost-effective action plans to reduce the levels of $SO_2$ and PM in several cities.

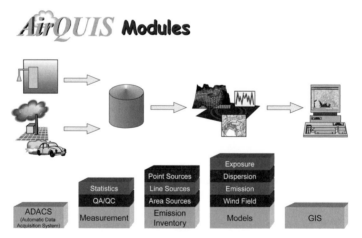

**Fig. 9.10.** Overview of module contents in AirQUIS

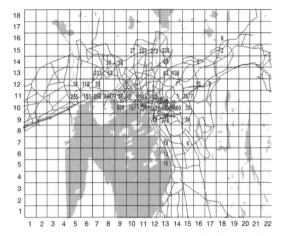

**Fig. 9.11.** Number of people exposed above national AQ Limit value for $NO_2$ (hourly) in 2001, calculated per $km^2$ grid cell

## GAMES and AQUAS: two Comprehensive Modelling and Decision Support Systems for Photochemical Pollution Control in Metropolitan areas

*Developer*: The Electronics for Automation Department group (University of Brescia - Italy), in cooperation with CESI and AGIP (Milano - Italy)

*Provider*: Contact/more information: Prof. Giovanna Finzi

*Application areas:*

The GAMES (Gas Aerosol Modelling Evaluation System) system has been designed:
- to analyse regional photochemical and aerosols episodes;
- to assess seasonal photochemical modelling simulations;
- to evaluate and select emission abatement strategies;
- to estimate the effectiveness of depollution measures in cost–effective analysis.

The AQUAS (Air QUality Alarm System) system can provide short-term urban pollution forecasts, by means of different modelling methodologies:
- non-stationary / non-linear grey box models,
- neural network,
- fuzzy logic,
- neuro-fuzzy models.

*Structural elements included/other main features:*

The GAMES system (Decanini and Volta 2002, Decanini et al. 2002) consists of some main modules:

- The transport and chemical model TCAM implements an accurate advection-diffusion scheme in terrain following coordinates with variable vertical spacing and a resistance-based dry deposition algorithm taking into account pollutant properties, local meteorology and terrain features. The chemical module has been designed by means of the Flexible Chemical Mechanism and has been updated with COCOH-97 mechanism. The model also includes the aerosol module (MAPS) allowing for size resolved and chemical split representation of particulate matter. The module includes gas to particle processes such as condensation and nucleation changing the total size and number concentration of the aerosol population. The model takes into account various aerosol phase compounds, such as sulphate, nitrate, ammonium, water, chloride, sodium, elemental carbon, primary organic compounds and 8 classes of secondary organic compounds.
- The meteorological model CALMET consists both of a diagnostic wind field module and micrometeorological modules for over water and overland boundary layers.
- The emission evaluation model POEM-PM provides present and alternative emission field of gas and PM compounds, estimated by means of an integrated *top-down* and *bottom-up* approach. The POEM-PM estimates NMVOC lumping according to chemical mechanism and size-chemical split of particulate matter emission fields.

The AQUAS system (Finzi and Volta 2001) implements two feedback loops. Air quality status forecast, given by the daily model, is supplied to the Authority in order to support decisions relevant to emission abatement strategies. These measures can effectively prevent smog episodes if they are planned enough time ahead. The designed system includes a second feedback improving operational effectiveness in the short term (hours). The information provided both by the hourly predictor and by the on-line meteo-chemical network allows the Air Quality Manager to monitor the current pollutant evolution. Any threshold value exceedance, not correctly forecast in advance, can be quickly recognised by the alarm system in order to apply short-term pollution control measures. The system also includes a tool to evaluate the performance of different predictors to foresee if the urban pollutants concentration will overcome an assigned threshold in terms of skill parameters, defined by the European Environment Agency.

*Technical platform:*

The GAMES system is operated from a PC platform with minimum requirements:
- PC Pentium III
- 256 MB RAM
- 20 GB hard disk
- LINUX operating system

The AQUAS system is operated from a PC platform with minimum requirements:
- PC Pentium II
- 256 MB RAM
- 5 GB hard disk

– Windows98/2000/ME operating system

*Developments during SATURN:*

Both systems have been designed and implemented during the project.

*Applications of the modelling and decision support system over Northern Italy:*

The GAMES system has been used
- to simulate pollution transport and chemistry.
  Modelling exercises have been performed by applying the GAMES modelling system over a domain of 240×232 km$^2$ with a horizontal grid size of 4km:
  – 5-7 June 1996; 1-6 June 1998; April-September 1996;
- to investigate the relations between aerosols and photochemical pollution:
  – 1-6 June 1998;
- to assess the effect of different emission strategies:
  – the impact of the implementation of EU directives on traffic emission scheduled up to 2005.
  – the impact of transport infrastructure strategies over urban areas.
- to evaluate the chemical regime by means of chemical indicators.
- by means of the simulation results, to identify simplified pollutant-precursor models, in order to integrate them in a multi-objective mathematical program, together with the estimate of the emission reduction costs.

The AQUAS system has been used to provide:
- hourly and daily ozone NO$_2$ and CO concentrations forecasts;
- maximum expected hourly ozone concentration value one day in advance;
- at noon maximum expected hourly ozone concentration value for the afternoon during the day itself for Milano, Brescia, Catania and Siracusa, processing 1994-2001 meteo-chemical data.

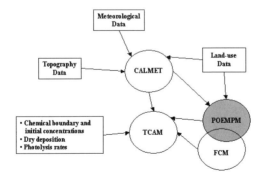

**Fig. 9.12.** GAMES structure

## Integrated urban AQM system (IUAQMS)

*Developer*: Department of Environment and Planning, University of Aveiro.

*Provider*: Contact/more information:  Prof. Carlos Borrego
http://www.dao.ua.pt/gemac

*Application areas:*

The system is a useful and friendly tool for air quality management in urban areas. The application aims cover:

- local scale air quality forecast for regulatory purpose including evaluation of legislation exceedance compliance;
- air quality impact assessment study;
- evaluation of effects on air quality resulting from future development scenarios;
- decision support tool for traffic planning and pollution control;
- providing information to estimate sustainability indicators.

*Structural elements included/other main features*:

- Data base of spatial information in GIS format including administrative maps, terrain elevation, land use, road network, traffic counting points and building volumetry;
- Data base of alpha-numerical information including historical meteorology and air quality data, emission inventory, census and other statistics;
- On-line measurements of traffic counting, meteorological and air quality parameters;
- Transportation model (VISUM) applied for estimation of traffic fluxes, for each city network road link, based on O-D matrixes;
- Transport emission model specially adapted for line sources (TREM) based on MEET/COST methodology which distinguishes between different vehicle type, technology and engine capacity;
- A boundary layer flow model coupled with a Lagrangean dispersion model (VADIS), specially developed and adapted to simulate urban air pollution in city centres, mainly oriented towards the street canyon pollutant dispersion;
- Geographic Information System (ArcInfo, ArcView) for data storage and processing;

*Technical platform:*

The system is operated from a PC server platform with the following minimum requirements:
- PC Pentium III - 1GHz
- 256MB RAM
- 1 GB Disk space
- 32MB Graphics card
- CD-R
- Windows 98/NT/2000

*Developments during SATURN:*

- Improvement of VADIS model to better describe obstacles (multi-obstacles) and flow fields (time varying flow fields), as well as emissions (multi-source, time varying);
- Development of interconnection between the transportation and emission models in order to provide hot and cold-start emission data with high spatial resolution;
- Development of interconnection between the emission and dispersion models;
- User-friendly interface for the modelling tools;

*Applications of the integrated urban AQM system:*

The integrated system has been applied to different cities under SUTRA research project (EVK4-CT-1999-00013, EU 5th FP) including Lisbon, Gdansk, Geneva, Genoa, Thessaloniki and Tel-Aviv in order to assess the air quality in the cities hotspots and evaluate different transportation scenarios.

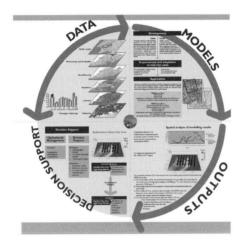

**Fig. 9.13.** Application of the integrated urban AQM system as a decision support tool

## IUEMIS: Integrated urban environmental and information systems

*Developer:* AUT/LHTEE, Environmental Informatics group

*Provider* Contact/more information: Dr. Kostas Karatzas

*This is a set of self-sustained modules that can be linked together, operate separately as stand-alone applications or be added to existing systems.*

*Application areas:*

- Environmental Information portals: web servers hosting air quality data, tools and services, in a customisable fashion
- Citizen-centred air quality information services via Internet (e-mail, web-based newsletter), mobile phone (SMS, WAP), PDAs (mobile Internet).
- Telematics systems for Air Quality Monitoring and Management: use of wireless telecommunications and mobile phone technologies for collecting, processing and managing-disseminating air quality information.
- Multimedia-based air quality presentations.
- Short-term air quality forecasting with the aid of statistical (CART, ARIMA, Regression) and Artificial Intelligence (Neural Networks) techniques.
- Web-based air quality modelling: use of sophisticated 3D air quality models via Internet, performing remote runs supported by a library of meteorological and emission scenarios. The Java-based user interface supports various air quality management post-processing tasks.

*Structural elements included/other main features*:

Full graphic interface modules operate and are made available to the "client" via Internet (http-based). Open source web-based GIS solutions are supported. Modules for the analysis of air quality time-series included. Client-server applications are supported. Information can be tailored to various user groups: experts, decision makers, general public. The content of electronic environmental information services and its delivery is supported.

*Technical platform:*

All modules can be made available remotely (i.e. no on-site installation is needed) if an Internet connection is assured. Alternatively, a moderate PC is sufficient. All modules are light applications, based on open-source software, and usually developed under Linux/Unix, while they can also be made available under Windows.

*Developments during SATURN:*

1. Theoretical Investigation of Urban Air Quality Management Systems Performance Towards Simplified Strategic Environmental Planning: the results revealed that the use of a scenario-based analysis should be limited to short-term, emergency based management, while for long term environmental management other means like the target-based approach are more suitable.
2. Development of a Requirements analysis methodology regarding Integrated Urban Air Quality Management and Information Systems: this is a method based on the use of questionnaires and the AHP method that can (and should) be used in the design of all Urban Environmental Management systems.
3. A Multimedia presentation and guidance related to environmental impact assessment information.
4. Development of an Internet-based air quality statistical forecasting module that uses CART, ARIMA, Regression Analysis and Neural Networks.
5. Development of an SMS and WAP module for the communication of environmental information via mobile phones.

6. Development of an e-mail and a Newsletter application for the web-based communication of environmental information.
7. Investigation of air quality data series structures with the aid of Fourier Analysis.
8. Investigation of atmospheric urban environment characteristics from antiquity until today.
9. Development of a ready to be used tool for the teletransmission and web-based dissemination of environmental and energy data.
10. Results on the investigation on citizen-centred information dissemination on multimodal information channels and GIS.

*Applications of the IUEMIS modules:*

Applications have been made in the frame of the following research projects and related sites-case studies: ECOSIM and IRENIE (Athens, Greece), DESPOTIS (Thessaloniki, Greece), AIR-EIA (at a country level for Austria, Germany and Greece), APNEE/APNEE-TU (Athens & Thessaloniki, Greece; Marseille, France; Madrid, Andalusia & Canary Islands, Spain; Oslo & Grenland, Norway; Dresden & Stuttgart, Germany).

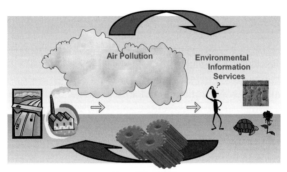

**Fig. 9.14.** The Environmental Information loop: balancing dissemination and action

## OPANA:OPerational Atmospheric Numerical pollution model for urban and regional Areas

*Developer*: Environmental Software and Modelling Group, Computer Science School, Technical University of Madrid (UPM)

*Provider*: Contact/more information: Prof. Roberto San José
http://artico.lma.fi.upm.es.

*Application areas:*

OPANA model is a representative of the Second Generation of Air Quality Models based on three-dimensional numerical solution of the Navier-Stokes partial differential equation system. This model was developed in the second half of the 90's. It is formed by a Visualization and Managing Interface of the Meteorological and Chemical modules. The MEMO model (developed at the U. of Karlsruhe) is the meteorological model used and the SMVGEAR module (University of Los Angeles, 1994) into the MEMO model in on-line mode. Currently the meteorological model is being replaced by the MM5 model (PSU/NCAR) the chemical module is being replaced by CMAQ model (Community Multi-scale Air Quality Modelling System, Models-3, U.S. EPA) incorporating aqueous chemistry, several aerosol dynamical modes and cloud chemistry. A sub-development of OPANA is the TEAP tool (A tool to evaluate the air quality impact of industrial emissions).

The OPANA model Version 5.0 (MM5-CMAQ) is suitable for:
- Assessments of present and future urban, regional and continental air quality disaggregated in space and time.
- Cloud chemistry, aerosol dynamical modes and aqueous chemistry is available.
- Short-term and long term forecasts.
- Planning tool: development of strategies.
- The meteorological information is included as part of the system.

*Structural elements included/other main features*:

OPANA model (Ver 5.0 (MM5-CMAQ)) incorporates the following capabilities:
- EMIMO and EMIMA models (emission inventory generators). EMIMO is a top-down emission model (developed in C) for general purpose (continental, regional and urban areas worldwide) and EMIMA is a bottom-up emission model developed for Madrid area.
- GRASS/GIS capabilities.
- VIS5D/DISAD capabilities.
- OCTAVE, PAVE and GRADS tools incorporated.
- Tcl-/TK environment
- Air pollution time series.
- In TEAP analysis system incorporated.
- Since version 2.0, OPANA is fully linked to the Internet.
- A special OPANA version is offered to the user only by Internet, providing the air quality forecasts produced in the UPM Labs. In this case no HW is required (except the Internet connection).

*Technical platform:*

Off-line version (this version is operated by the user only):
- Ver 5.0 (MM5-CMAQ) is mounted over LINUX PC Red Hat 3.7
- 1 GB RAM
- 120 GB HD.
- 2 PC's (one PC for MM5 and the other one for CMAQ).
- Internet connection.

Note: TEAP can require up to 12 – 16 PC's depending on the user requirements (industrial plant).

On-line version. This version assumes that the UPM Lab. operates the model and offers the results to the user through the Internet. In includes a warning system to the user (mobile telephone SMS, e-mail) based on the air quality forecasts (exceedances, pollution index, etc.) fully customised to the user.

*Developments during SATURN:*

In SATURN we have developed all the evolution from the use of MEMO model and SMVGEAR to MM5-CMAQ (SMVGEAR is included in CMAQ but also RADMS chemical mechanism).

*Applications of the OPANA system:*

OPANA model is operationally implemented at:
- City of Bilbao (Spain) (EQUAL, IST) project (2001).
- City of Leicester (United Kingdom) (EQUAL, IST) project (2001). In this case an on-line version continue to be operating for Leicester City Council under private contract (2001-2003).
- Asturias Regional Government (Spain) (1997-1999). A nested version was installed in 1998 in Oviedo (Spain) with an IBM-RISC 6000 workstation. It is operated by the user.
- Madrid Community (Spain) (EMMA, IST) project (1996-1998). Operated by the user.
- Madrid City: Two versions 1) (EMMA, IST) project (1996-1998). 2) Private contract (2002) (MEMO-SMVGEAR).
- Quito (Equador). Private contract (World Bank) (1999-2002).
- Las Palmas de Gran Canaria (Canary Islands, Spain) (2002-2004) (MM5-CMAQ version). Private contract.
- Andalusia region (Spain). APNEE-TU, IST project (2002-2003).
- Canary Islands region (Spain). APNEE-TU, IST project (2002-2003).

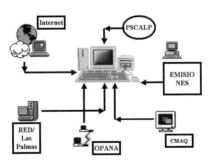

**Fig. 9.15.** Scheme of the future application of the OPANA system in Las Palmas City in Canary Islands (Spain).

## OPUS-AIR: Integrated assessment policy system for European Urban air pollution studies

*Developer*: AUT/LHTEE Thessaloniki group

*Provider*: Contact/more information: Prof. Nicolas Moussiopoulos

*Application areas:*

The OPUS-AIR system evolved from the ZEUS model system and its forerunners and is a policy oriented system for the integrated assessment of technical and non-technical measures that are put forward in order to reduce urban air pollution levels. In contrast to the majority of current air pollution model systems, OPUS-AIR not only assesses the impact of air pollution abatement strategies in terms of their influence on air quality, but also takes into account the economic impacts stemming from their implementation. Thus, OPUS-AIR evaluates the total costs required for the application of any proposed strategy, treating in parallel the environment as an economic commodity by quantifying the costs of any form of environmental degradation caused by air pollution. Thereby, emission control measures are assessed through the combined application of advanced air quality models and validated economic evaluation methods.

Analyses with OPUS-AIR are performed through a multiple-step procedure that involves:

- built-up of detailed emission inventories for the areas examined,
- construction of cost-curves for selected pollutants along with the estimation of the total costs required for the implementation of each of the strategies proposed,
- application of either simple or more sophisticated air quality models (e.g. the MEMO/OFIS cascade, the EZM system or the multi-scale model cascade ZEUS),
- evaluation of the economic benefits on account of air quality improvement with regard to human health and human productivity,
- graphical representation of results and air quality indicators on a GIS platform.

With OPUS-AIR the assessment of the proposed strategies is conducted a) through the comparison of the current air pollution levels with the concentration levels resulting after the hypothetical implementation of abatement measures and b) having as an additional criterion the minimisation of the total costs required for or arising from the adoption of the measures.

Based on the multi-pollutant, multi-effect concept, OPUS-AIR aims at providing policy-makers with a reliable tool for the objective assessment of the most cost-effective packages of measures, the latter being allocated according to the particular features and needs of the area examined.

*Structural elements included/other main features*:

OPUS-AIR composes of ZEUS and an Economic Assessment Module (EAM). ZEUS is comprised of different scale models with appropriate interfaces between them:

- The EZM system which includes mesoscale model MEMO for the prognosis of the mesoscale meteorological conditions in the area of application coupled with the chemical model MARS providing pollutant concentration information.
- The microscale domain - resolving the geometrical details of a part of the city - is allocated within the finest mesoscale domain. Quasi-steady MIMO computations provide hourly averaged wind and pollutant concentration fields.
- The mesoscale model provides the microscale model with the appropriate initial and boundary conditions through a one-way coupling technique.

*Technical platform:*

It can be operated in any UNIX environment.

*Developments during SATURN:*

Development and improvements of the system are on-going. Partial goals of current work are the development of (i) a canopy scale model, (ii) a method for the coupled treatment of wind flow and pollutant dispersion, (iii) a "top-down/bottom-up" model cascade strategy including advanced turbulence models and (iv) appropriate modules for the description of condensation and heterogeneous chemistry in ZEUS (v) Introduction of Multiple Objective Linear Programming (M.O.L.P.) for selecting optimised intervention bundles in EAM.

*Applications of the OPUS-AIR system:*

OPUS-AIR was applied to the Greater Athens area (GAA) for evaluating the impact of various air quality regulations concerning technical measures for the re-

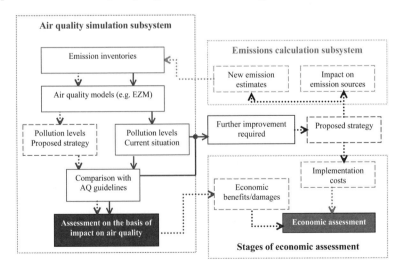

**Fig. 9.16.** Flow diagram of the OPUS-AIR assessment system

duction of $NO_x$ and NMVOC emissions from the most important emission sources of the area.

The MEMO & MARS system has been applied to several urban areas (Greater Milan Area, Greater Athens Area, Greater Stuttgart Area, etc.) in order to predict meteorological conditions favouring the accumulation of various pollutants (e.g. photochemical smog), while MIMO has been applied to particular street canyons, e.g. Podbielski Strasse and Goettinger Strasse in Hanover, Germany.

## Photosmog pollution episode warning system

*Developer*: Department of Human Exposure Research and Epidemiology at UFZ-Centre for Environmental Research Leipzig-Halle Ltd., Germany.

*Provider*: Contact/more information: Prof. O. Herbarth, Dr. U. Schlink
http://www.ufz.de/spb/expo/index.html

*Application areas:*

Short-term forecasting of $O_3$ pollution based on statistical models. An important aim of the developed statistical models was a comparative assessment of different statistical forecasting techniques. Another significant contribution is the assessment of the health effects of $O_3$ pollution. The identified associations between $O_3$ concentration and respiratory symptoms enable to quantify the health risk occurring during an air pollution situation.

*Structural elements included/other main features:*

For $O_3$ forecasting several time series models have been developed, such as autoregressive models, regression models, transfer-function models, and component models (based on the Kalman filter). These linear models have been complemented by a non-linear approach. The latter was applied to testing $O_3$ time-series for non-linear dynamics.

*Technical platform:*

All developed modules operate on a personal computer and are based on software for statistical modelling: Matlab, GenStat, Statistica

*Developments during SATURN:*

Work within SATURN was focused on the development of the transfer-function model and the application of several statistical models with data of different European regions. In co-operation with partners in Thessaloniki, Brescia and Budapest we studied the performance of statistical model with their data. Another contribution to SATURN was the analysis of peculiarities of the $O_3$ time-series and their interpretation in terms of local meteorological phenomena.

*Applications of the episode warning system:*

The statistical models have been mainly applied in intercomparison studies. In co-operation with the EU project APPETISE (5<sup>th</sup> framework) the forecasting performance of each of the statistical approaches was assessed and recommendations were made for their application in operational air quality forecasting.

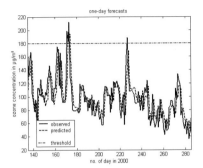
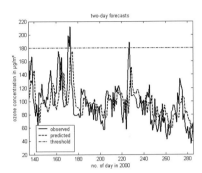

**Fig. 9.17.** Daily maxima of 1 hr $O_3$ observations in Berlin; day No. 1 is the 1/1/00. The dashed line represents the predictions of the transfer-function model based on temperature and $NO_x$ concentration as input. The horizontal line is the threshold defining the exceedances

## UDM: Urban Dispersion and exposure Modelling system

*Developer*: Finnish Meteorological Institute (FMI) group in Helsinki

*Provider*: Contact/more information: Dr. Jaakko Kukkonen and Dr. Ari Karppinen

*Application areas:*

The system is suitable for:
– the assessment of traffic flows, emissions originated from mobile and stationary sources, urban air quality, and the exposure of population to air pollution, and their spatial and temporal variations. The structure of the modelling system has been described by Karppinen et al. (2000a) and its evaluation against the data from an urban air quality monitoring network by Karppinen et al. (2000b) and Kousa et al. (2001).
– short-term forecasts of urban air pollution, when coupled to the meteorological forecasting model HIRLAM.
– planning tool: can be used for evaluating, e.g., the effect of various abatement strategies, and future land use and traffic planning scenarios.
– air quality assessments in cities on a routine basis

*Structural elements included/other main features:*

- Includes modules for evaluating traffic flows, emissions (including also aviation and marine traffic), urban scale dispersion models that include point and line source submodels (both roadside environments and street canyons; Kukkonen et al. 2001a), and a detailed population exposure model (Kousa et al. 2002).
- The graphical output of the system is GIS-based (MapInfo).
- The post-processing module enables also the statistical analysis of sequential hourly time series, and the creation of two-dimensional maps (values in a receptor grid and concentration isolines).
- The system includes a road network emission and dispersion model (CAR-FMI), which can be applied as a stand-alone user-friendly PC program. This model includes a Windows-based user interface and a user's manual in English, as well as an option to present results easily in a GIS system. The programme can be requested from the responsible persons, to be utilised for research purposes. This model has been evaluated against the data of two roadside field campaigns (Kukkonen et al. 2001b; Öttl et al. 2001).

*Technical platform:*

The whole system can be operated on the FMI mainframe and a supercomputer, except for the traffic planning module that is operated by the Helsinki Metropolitan Area Council (YTV). A version of the CAR-FMI roadside dispersion model can also be used on a PC.

*Developments during SATURN:*

The modelling system has been evaluated against the data from an urban measurement network (Karppinen et al. 2000b; Kousa et al. 2001). The individual modules of the system have also been evaluated against the data of three field measurement campaigns in cooperation with YTV, NERI (Denmark) and Graz University of Technology (Austria) (Kukkonen et al. 2001a, b; Öttl et al. 2001). The system has been extended to treat $PM_{2.5}$ (Tiitta et al. 2002; Karppinen et al. 2002) and $PM_{10}$ (Kukkonen et al. 2001c). The population exposure model has also been developed within SATURN (Kousa et al. 2002).

*Applications of the urban dispersion and exposure modelling system:*

- The system has been utilised widely in urban air quality assessments in Finland, in more than ten cities. The system has also been used in a few other countries.
- The system is being used for AQ forecasting in several cities in Finland. The meteorological forecast is predicted by the HIRLAM model.

The system is continuously used in various research projects. For instance, it has been utilised for evaluating the effect of future land use and traffic planning scenarios in the Helsinki Metropolitan Area.

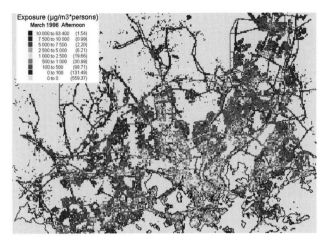

**Fig. 9.18.** The predicted average exposure of the population to $NO_2$ concentrations in Helsinki Metropolitan Area ($\mu$g m$^{-3}$ persons), evaluated for the afternoon period, as an average value in March '96. The value in brackets refers to the area (km$^2$) with the exposure in the selected range. The grid size is 100m × 100m and that of the depicted area 25 km × 34 km.

## URBIS: Urban Information and management System

*Developer*: TNO-MEP, P.O Box 243, 7300 AH Apeldoorn, The Netherlands.

*Provider*: Contact/more information: Prof. Dick van den Hout,
hout@mep.tno.nl

*Application areas:*

- Municipalities and individual urban districts
- Detailed assessment and mapping of the air quality and noise emissions and levels, based on high resolution modelling
- Prognosis of future air quality and noise level
- Analysis of the effectiveness of possible measures
- Calculation of potential exposure of population and mapping of health risks and annoyance.

*Structural elements included/other main features*:

The system combines the regular databases and modelling tools available in the Netherlands. It uses detailed databases for all sources in the city. Concentrations that are due to elevated sources are calculated with a standard Gaussian plume model. This model is also used for near-ground sources, but for those cases the microscale fine structure of the concentration pattern is calculated by a microscale model, in particular the CAR model for streets. The system addresses the needs brought into life by the new EU air quality directives, whose limit values need to be complied with throughout the year and throughout the territory of agglomerations.

*Technical platform:*

PC based; partly based on GIS. In 2002, a stand-alone version is being developed.

*Developments during SATURN:*

URBIS has been developed in SATURN. Several challenges had to be faced in the development of the system. In the first place, the individual models, in particular the street traffic model CAR needed to be modified so as to give results that could be combined with the results of the other models, but it had on the other hand to remain as consistent as possible with the original model. This required not only trivial format changes but also conceptual changes. In particular, the CAR model has a low longitudinal and arbitrarily high transversal resolution and is undefined at street intersections. Secondly, the high resolution demanded a carefully designed strategy for distributing receptors in the city. Several types of receptors were used to enhance the resolution at locations of high concentration gradients: in addition to a basic set of rectangular receptors, additional sets of receptors were defined placed relative to the position of low local sources and relative to both buildings in combination with streets respectively. Thirdly, emission data for medium and small enterprises had to be designed, as these data were generally not available. A tiered approach was defined, starting with national (and if available local) emission data, followed by estimates based on local activity data related to an economic classification and, finally, more detailed inventory work for the most relevant sources.

These methods were not developed in close relation with other developments. A close link was maintained with similar development work in the field of noise, with which a fully integrated system was built, and with the field of external safety, where until now only a link on the level of data exchange was deemed useful. Also, emphasis was placed on the presentation of the results and the use for the various authorities. In interviews and workshops it was found that there was a wide range of potential uses, and also a wide range of views by users on what was useful to have available. Among the most important applications were the generation of a general overview of the environmental quality in the city, the identification of limit value exceedances and scenario analysis. In the last stage within SATURN, the wish of some authorities to have a stand-alone version is being addressed.

*Applications of the URBIS system:*

URBIS has been applied in various cities and districts in the Netherlands, in some cases for both air quality and noise, in other cases for air quality or noise alone. Currently it is being installed as a software tool in Rotterdam.

# Chapter 10: Conclusions

N. Moussiopoulos

Aristotle University Thessaloniki, Greece

## 10.1 Achievements of SATURN

The major scientific achievement of SATURN is its contribution to a better understanding of physical processes occurring in the urban atmosphere. Such processes are nowadays evident in most numerical simulation models and an extensive database of laboratory experiments exists within which these processes are present. In particular, research in SATURN led to an improved scientific knowledge of the spatial and temporal characteristics of urban air pollution in Europe. The associated investigations led to the identification of critical meteorological conditions responsible for severe air pollution episodes in conjunction with the specification of the regions where these episodes occur. Emphasis was put on the study of circulations that are associated with low wind conditions and thermal effects. Towards this aim, high-quality experiments were performed and modelling tools were improved and developed, including novel multi-scale model cascades.

Concerning air pollutant emissions, a harmonised method for compiling urban emission inventories was developed in SATURN. Guidelines developed include the methodology requirements on data resolution and emission factors. A protocol for qualitative and quantitative comparisons of emission inventories was also established. Individual studies confirmed that suitable combinations of $NO_x$ and VOC emission reductions are needed for an effect-oriented and cost-effective strategy aiming at the reduction of ozone. For this reason, special emphasis was put on assessing the effectiveness of ozone precursor emission reductions at the local and regional scales.

Field experiments led to a better insight into the characteristics of polluted urban air. There was special focus on the particulate pollution, including:

- the assessment of contributions from various pollution sources due to human activities (e.g. traffic, industry) and biogenic emissions (e.g. sea salt aerosols);

- a better understanding of the particulate matter characteristics (size distribution, chemical composition, transport properties etc); and
- an improved knowledge on important processes such as the suspension of particles from deposited road dust.

Specifically, it was found that the $PM_{10}$ mass and most of the analysed species are dominated by the fine fraction. Traffic is the dominating source of ultrafine particles in busy streets, while also having a significant contribution to $PM_{10}$. The application of averaged PM data, collected continuously, in combination with routine monitoring data and manually counted traffic rates, was found to be a powerful tool to determine the contribution and emission factors of particles from diesel and petrol vehicles from the actual car fleet under normal driving conditions. The method may prove useful for demonstrating the effect of air pollution abatement measures.

Other experiments concentrated on photochemical pollution especially in South Europe revealing pathways that lead to secondary particle generation. Regarding the chemical composition, the coarse mode (composed of sea salts and nitrates) is most important at the beginning of the event, but decreases strongly when the pollution increases, to become much lower than the accumulation mode. The latter mode is totally dominated by sulphates and ammonium and the smaller particles of this mode become progressively more important during the event. The focus of other experiments was the investigation of the thermal characteristics of the urban atmosphere and their effects on the boundary layer flows and structure.

Improvement of receptor modelling techniques within SATURN has led to the more accurate estimation of source-receptor relationships for various pollution sources in urban areas and their surroundings. A better knowledge on the relative contribution of individual sources is a prerequisite for cost-effective control strategies.

Urban scale dispersion models were refined in the framework of SATURN. Efficient interfaces were developed for linking such models to suitable regional scale models. In addition, improved parameterisation methods were developed and numerical techniques were refined. Work included evaluating the aerosol behaviour in urban areas using detailed chemical and aerosol dynamical models. In addition, transport-chemistry models for the evaluation of emission reduction strategies, for providing information to the public, and as the central part of models for forecasting episodes and for the calculation of human exposure were developed. SATURN has shown a special interest in the issue of air pollution exposure, providing measurements and modelling systems that can be used in traffic and urban planning, in environmental management and in analysing the health effects of air pollution.

Both the concept and the application range of local-scale models progressed significantly within the framework of SATURN. Applications included simulations of the air motion, turbulent field and heat fluxes close to building walls, as well as their effect on pollutant dispersion. The investigations of several SATURN groups on the effects of buildings on the developed airflow within the street have shown the essential role of the street and building geometry on the ventilation and air pollution dispersion from the street. Vehicle induced turbulence within the street corresponds to an important influence of traffic on the dispersion characteristics. The numerical results show the importance of the measuring position when models are to be evaluated thus suggesting the significance of an adequate siting of urban monitoring stations. Computational fluid Dynamical (CFD) modelling may prove useful for recommendations regarding suitable monitoring sites.

SATURN led to significant progress of the state-of-the-art with regard to model quality assurance. A straightforward procedure for evaluating local scale models has been formulated based on discussions between numerical model developers and experimentalists. It is characteristic of the urban canopy layer that the velocity and concentration fluctuations are larger than the corresponding mean values causing a large inherent uncertainty in the data, and this needs to be quantified before the data can be used for validation purposes. A methodology for the quantification was developed and applied. On the other hand, wind-tunnel experiments have been carried out to quantify how numerical model results depend on domain size and geometrical resolution. Furthermore, valuable new knowledge regarding quality assurance of urban scale models resulted from intercomparison and validation activities, an example for the latter category being ESCOMPTE_INT, an exercise based on the ESCOMPTE pre-campaign data sets.

Work performed in SATURN elucidates the great potential of remote sensing with regard to urban air quality management. Apart from the demonstration of monitoring approaches based on remote sensing, it was shown that interferometry may contribute to the better representation of terrain elevation and land use and the provision of reliable information on roughness and building configurations.

Knowledge and tools acquired in the framework of SATURN were integrated in order to make them directly suitable for applications related to environmental policy and to support urban air quality management. Gradually, the integrated tools for modelling and predicting air pollution improve in quality and efficiency, while novel telematics techniques are being applied for informing the public on air pollution. Moreover, methods for the air quality forecasting have been developed and improved with special emphasis to those concerning ozone forecast.

The policy relevance of the above scientific achievements is obvious, given their direct influence on the formulation of improved tools for urban air quality assessments. Hence, in the last few years new methods to determine the contributions of various sources to air pollution in conurbations were developed and existing ones were improved. Such methods may be utilised by urban authorities wishing to have insight into the possibilities of reducing air pollution levels or in controlling anticipated increases of those levels. Furthermore, methods were refined for the prediction of the effect of long-term emission changes. Such methods may considerably help formulating and evaluating air pollution abatement strategies.

## 10.2 Remaining gaps of knowledge

Despite the scientific progress achieved in SATURN, several issues related to urban air pollution have not yet been fully explored. Examples are photochemical air pollution in South European cities and particulate matter in conurbations. Moreover, more effort is needed to arrive at reliable hotspot assessments. This refers to both local scale emission estimates and appropriate multi-scale dispersion modelling approaches. In particular, the following appear to be the remaining gaps of knowledge:

- Pollutant emissions due to real traffic behaviour. Much more detailed information is required for delivering appropriate emission estimates for local scale problems, typically investigations of traffic flows between crossings, traffic signals, etc. Research should focus on the description of the actual vehicle fleet, a better knowledge on the engine working conditions and obtaining access to instantaneous emission data.

- Sources of $PM_{10}$ and $PM_{2.5}$ affecting the urban atmosphere. Knowledge of local emissions and regional transport of $PM_{10}$ and $PM_{2.5}$ is limited and associated emission inventories are at an early stage of development. The formation and regional transport of secondary particles are so far only poorly understood. Such mechanisms are inadequately treated in present modules and the few models that include the description of particle formation and transport were not yet properly evaluated and tested.

- Physico-chemical properties of urban aerosol. More knowledge is needed regarding the chemical and physical properties of the particles (EC, OC, PAH, metals etc.), e.g. surface properties, volatility, hygroscopicity, morphology. For this purpose it is necessary to apply advanced analytical techniques. Also the process of nano-particle formation has not been unambiguously determined. In this context it is recommended to establish a systematic urban network in Europe for the monitoring of fine particles (including their size distribution and chemical composition) in order to achieve a more detailed and region-depending picture of the origin and potential risks associated with particles.

- Assessing the impact of urban sources vs. long-range transport with regard to urban air quality. Dispersion and transformation processes at the local scale differ considerably from those at the urban and regional scales. Future research should concentrate on novel simulation methods based on nested domains with different models in order to accurately predict to what extent air quality in various city parts is affected by sources within the city boundary or can be attributed to long-range transport.

- Prediction of pollutant concentrations at hotspot locations. Further research is needed for understanding processes taking place during the residence of pollutants in the urban canopy. Furthermore, more effort is necessary for better simulating local phenomena observed in the vicinity of buildings, such as thermal convection due to the heating of the ground and walls and traffic induced influences.

- Disparity between sophisticated models and tools needed for legislative purposes. Obtaining reliable predictions of concentration fields inside the urban canopy layer is not straightforward. The more sophisticated CFD codes available for that purpose are still relatively new and mostly in their development phase. The fitness for purpose of these models was demonstrated at the example of rather idealised cases only. The prediction of concentration levels for detecting exceedances of limit values is hardly possible at the temporal and spatial resolution required by the existing air quality legislation. The simple models, on the other hand, lack sufficient physical foundation. Targeted efforts are needed to close the gap, primarily towards adapting scientific models to regulatory needs.

- Quantitative level of confidence of model predictions. The widespread use of urban scale models in various countries in Europe requires a comparable increase of co-ordinated international initiatives aiming at the verification of model quality, model evaluation and setting standards on their performance. A standardised quality assurance procedure would give clear guidance to developers and users of urban air quality models as to how to properly assure the quality of their models and results. In this respect, activities towards model intercomparison and validation may prove extremely helpful.

- Need of data in support of model quality assurance work. SATURN started to provide suitable and quality-approved data for model validation purposes in a convenient and generally accessible form. In general, however, such data refer to rather idealised cases with simple geometries. In parallel to the increasing potential of numerical models, more and higher quality data will be required for meeting the needs of complete evaluation procedures.

- Integrated assessment framework. A robust methodological framework for integrated assessment at the local and urban scales is still lacking. Specifically, the linking between traffic modelling, emissions modelling, dispersion modelling and personal or community exposure modelling could be regarded as weak. Also the combined and fully integrated use of measuring and modelling for assessing air quality is still in its infancy. The associated disciplines tend so far to act independently.

# References

## Chapter 1

URL 1.1: http://aix.meng.auth.gr/saturn/

## Chapter 2

Adams HS, Niewenhuijsen MJ, Colville RN (2001) Determinants of fine particle (PM$_{2.5}$) personal exposure levels in transport microenvironments. London, UK, Atmos Environ 35: 4557–4566

Almbauer RA, Öttl D, Bacher M, Sturm PJ (2000) Simulation of the air quality during a field study for the city of Graz. Atmos Environ 26B, No 3: 379–390

Britter RE, Hanna S (2003) Flow and dispersion in urban areas. Annu Rev Fluid Mech 35: 469–496

Brown MJ (2000) Urban parameterisations for mesoscale meteorological models. In: Boybeyi Z (ed), Mesoscale Atmospheric Dispersion, Southhampton, UK, WIT Press, Ch 5, pp 193–255

Grimmond CSB, Oke TR (2002) Turbulent heat fluxes in urban areas: observations and a local-scale urban meteorological parameterisation scheme. J. Appl. Met., in press

Hanna SR, Britter RE (2002) Wind flow and Vapor Cloud Dispersion at Industrial and Urban Sites. New York: Am. Inst. Chem. Eng.

Hosker RP(1985) Flow around isolated structures and building clusters: a review. ASHRAE Trans 91: 1671–1692

McInnes G (1996) CORINAIR Atmospheric Emission Inventory Guidebook, European Environment Agency

Oke TR (1987) Boundary Layer Climates. London, UK, Routledge

Macdonald R, (2000) Modelling the mean velocity profile in the urban canopy layer. Bound-Lay Meteorol 97: 25–45

Moussiopoulos N, Sahm P, Kessler Ch (1995) Numerical simulation of photochemical smog formation in Athens, Greece-a case study. Atmos Environ 29: 3619–3632

Pavageau M, Rafailidis S, Schatzmann M (2001) A comprehensive experimental databank for the verification of urban car emission dispersion models. Int J Environ Pollut 15: 417–425

Scaperdas A, Colville R (1999) Assessing the representativeness of of monitoring data from an urban intersection site. Atmos Environ 33: 661-674

Schatzmann M, Leitl B (2002) Validation and application of obstacle resolving urban dispersion models. Atmos Environ 36: 4811–4821

Schlunzen KH (2002) Simulation of transport and chemical transformation in the atmospheric boundary layer- a review on the past 20 years developments in science and practice. Meteorol Z, in press

Wåhlin P, Palmgren F, van Dingenen R (2001) Pronounced decrease of ambient particle number emissions from diesel traffic in Denmark after reduction of the sulphur content in diesel fuel. Atmos Environ 35/21: 3549–3552

# Chapter 3

Catenacci G, Riva M, Volta M, Finzi G (2000) A Model for Emission Scenario Processing in Northern Italy. In: Borrell, PM, Borrell P (eds) Proc. EUROTRAC-2 Symposium '98, Witpress, Southampton, vol 2, pp 720–724

Cirillo M, de Lauretis R, del Ciello R (1996) (ENEA): Review study on European urban emission inventories(final draft).European Environment Agency, European topic center on air emission, Topic report 1996 Document EEA/96

Decanini E, Pola M, Polla Mattiot F, Volta M (2002) Application of REMSAD and GAMES modelling systems on a particulate matter and ozone episode in Milan metropolitan area. In: Sturm PJ (ed) Proc. 8th International Conference on Harmonisation within Atmospheric Dispersion Modelling for Regulatory Purposes, Sofia

EEA (1998) http://tiger.eea.eu.int/aegb/default.htm

Finzi G, Volta M, Carnevale C, Ragni C (2002) POEMPM: An emission model for size and chemical description of particulate matter. In: Midgley PM, Reuther M, Williams M (eds) Proc. EUROTRAC-2 Symposium 2002. Springer Verlag, Berlin

Friedrich R., et al. (2002) GENEMIS-2 final report EUROTRAC 2

Gabusi V, Volta M, Veraldi S, Veronesi V (2001) Road traffic impact on photochemical pollution: Brescia Metropolitan Area Case. In: Midgley PM, Reuther M, Williams M (eds) Proc. EUROTRAC-2 Symposium 2000. Springer Verlag, Berlin, pp 1146–1149

Kühlwein J, Friedrich R (1999) Uncertainties of modelling emissions from motor vehicles. In: Sturm PJ (ed) Proc. 8th International Symposium Transport and Air Pollution. Graz

Kühlwein J, Friedrich R (2000) Uncertainties of modelling emissions from motor vehicles. Atmos Environ 34, No 27: 4603–4610

McInnes G (ed) (1996) Atmospheric emission inventory guidebook. A joint EMEP/CORINAIR Production, EEA, Copenhagen, February 1996, pp B710/9–11

McInnes G (1996) CORINAIR Atmospheric emission inventory guidebook. EEA, Feb.1996

Mensink C (2000) Validation of urban emission inventories. Environ Monit Assess 65: 31–39

Ntziachristos L, Samaras Z (2000) COPERT-III (Computer Programme to Calculate Emissions from Road Transport) – Methodology and emission factors (version 2.1). Technical report No 49, European Environment Agency. http://vergina.eng.auth.gr/mech/lat/copert/copert.htm, p 86

Sturm PJ (1999) Investigations at the Institute for Internal Combustion engines and Thermodynamics, Graz University of Technology

Sturm PJ, Winiwarter W, ApSimon H, Böhler T, Lopes M, Borrego C, Mensink C, Blank P, Friedrich R, Volta M, Finzi G, Moussiopoulos N (1998) Harmonised method for the compilation of urban emission inventories for urban air modelling. (Final report of

SATURN WG Val3 and GENEMIS WG Emission Inventories for Model Validation Activities, Institute for Internal Combustion Engines and Thermodynamics), TU-Graz, Austria

Sturm PJ, Sudy Ch, Almbauer RA, Meinhart J (1999) Updated urban emission inventory with a high resolution in time and space for the city of Graz. Sci Total Environ 235: 111–118

De Vlieger I (1997) On-board emission and fuel consumption measurement campaign on petrol-driven passenger cars. Atmos Environ 31: 3753–3761

Vlaamse Milieumaatschappij (VMM) (2001) Lozingen in de lucht (Air emission report) 2000, Flemish Environmental Agency, Aalst (in Dutch)

Volta M, Finzi G (1999) Evaluation of EU road traffic emission abatement strategies in northern Italy by a photochemical modelling system, EUROTRAC Newsletter 21, pp 29–35

URL 3.1: http://www.aeat.co.uk/IMPRESAREO/index.htm

# Chapter 4

Adams HS, Kenny LC, Nieuwenhuijsen MJ, Colvile R, Gussman R (2001a) Design and validation of a high flow personal sampler for $PM_{2.5}$. J Exposure Anal Env Med 11: 5-11

Adams HS, Nieuwenhuijsen MJ, Colvile RN, McMullen MAJ, Khandelwal P (2001b) Fine particle (PM2.5) personal exposure levels in transport microenvironments, London, UK. Sci Total Environ 279: 29-44

Adams HS, Nieuwenhuijsen MJ, Colvile RN (2001c) Determinants of fine particle (PM2.5) personal exposure levels in transport microenvironments, London, UK. Atmos Environ 35: 4557-4566

Adams HS, Nieuwenhuijsen MJ, Colvile RN, Older MJ, Kendall M (2002) Assessment of road users' elemental carbon personal exposure levels, London, UK. Submitted to Atmos Environ, October 2001, Manuscript

Almbauer RA, Pucher K, Sturm PJ (1995) Air quality modeling for the city of Graz. Meteorol Atmos Phys 57: 31–42

Benson P (1992) A review of the development and application of the CALINE3 and 4 models. Atmos Environ 26B, No 3: 379–390

Berkowicz R, Ketzel M, Vachon G, Louka P, Rosant JM, Mestayer PG, Sini JF (2002) Examination of traffic pollution distribution in a street using the Nantes'99 experimental data and comparison with model results. Water Air Soil Poll: Focus 2, 311-324

Borrego C, Barros N, Valinhas MJ, Santos P, Miranda AI, Carvalho AC, Tchepel O, Lopes M, Abreu F (1998) Final Report AMAZOC (Atmospheric Environment in Coastal Zones: Assessment of Ecosystem Load Capacity - PRAXIS/3/3.2/AMB/38/94). Departamento de Ambiente e Ordenamento, Universidade de Aveiro, Portugal, 213 pp

Borrego C, Barros N, Barros N, Lopes M, Conceino M, Valinhas MJ, Tchepel O, Coutinho M, Lemos S (1999) Data collection for mesoscale models validation: a field campaign. In: Borrell, PM, Borrell P (eds) Proc EUROTRAC-2 Symposium'98, WIT Press, Southampton, pp.467-471

Etling D (1990) On Plume Meandering under Stable Stratification. Atmos Environ 8: 1979–1985

Granberg M, Niittymäki J, Karppinen A, Kukkonen J (2000) Combined application of traffic microsimulation and street canyon dispersion models, and evaluation of the modelling system against measured data. In: Sucharov L, Brebbia CA (eds) Urban Transport VI, Urban Transport and the Environment for the 21st Century. WIT Press, Southampton, pp 349–358

Karppinen A, Joffre SM, Kukkonen J, Bremer P (2001) Evaluation of inversion strengths and mixing heights during extremely stable atmospheric stratification, Int J Environ Pollut 16, Nos 1-6

Kukkonen J, Salmi T, Saari H, Konttinen M, Kartastenpää R (1999) Review of urban air quality in Finland. Boreal Environ Res 4, No 1: 55–65

Kukkonen J, Valkonen E, Walden J, Koskentalo T, Karppinen A, Berkowicz R, Kartastenpää R (2000) Measurements and modelling of air pollution in a street canyon in Helsinki. Environ Monit Assess 65 (1/2): 371–379

Kukkonen J, Valkonen E, Walden J, Koskentalo T, Aarnio P, Karppinen A, Berkowicz R, Kartastenpää R (2001a) A measurement campaign in a street canyon in Helsinki and comparison of results with predictions of the OSPM model. Atmos Environ 35, No 2: 231–243

Kukkonen J, Härkönen J, Walden J, Karppinen A, Lusa K (2001b) Evaluation of the dispersion model CAR-FMI against data from a measurement campaign near a major road. Atmos Environ 35, No 5: 949–960

Lazar R, Podesser A (1999) An urban climate analysis of Graz and its significance for urban planning in the tributary valleys east of Graz. Atmos Environ 33: 4195–4209

Lohmeyer A, Müller WJ, Bächlin W (2002) A comparison of street canyon concentration predictions by different modellers: final results now available from the Podbi-Exercise. Atmos Environ 36: 157-158

Louka P, Vachon G, Sini JF, Mestayer PG, Rosant JM (2002) Thermal effects on the airflow in a street canyon - Nantes '99 experimental results and model simulation. Water Air Soil Poll: Focus 2, No 5-6: 351-364

Mestayer PG, Durand P (2002) The UBL/CLU-Escompte experiment: description and first results. In: Proc. 4th symposium on Urban Climatology, May 2002, Norfolk, VA (Paper 3.1)

Öttl D, Kukkonen J, Almbauer RA, Sturm PJ, Pohjola M, Härkönen J (2001) Evaluation of a Gaussian and a Lagrangian model against a roadside dataset, with focus on low wind speed conditions. Atmos Environ 35: 2123–2132

Piringer M, Baumann K (1999) Modifications of a valley wind system by an urban area – experimental results. Meteorol Atmos Phys 71: 117–125

Vachon G (2001) Transfert des pollutants des sources fixes et mobiles dans la canopée urbaine: évaluation expérimentale. Doctoral thesis, University of Nantes, 23 October 2001

Vachon G, Rosant JM, Mestayer PG, Sini JF (1999) Measurements of dynamic and thermal field in a street canyon, URBCAP Nantes '99. In: 6th International Conference on Harmonisation within Atmospheric Dispersion Modelling for Regulatory Purposes. 11-14 Octobre 1999, Rouen, France, Volume of Preprints, pp 239–240

Vachon G, Rosant J-M, Mestayer PG, Louka P, Sini J-F (2000a) Pollutant Dispersion in an Urban Street Canyon in Nantes: Experimental Study. In: Proc. Symposium 2000 EUROTRAC-2, 27-31 March 2000, Garmisch-Partenkirchen

Vachon G, Rosant JM, Mestayer PG, Louka P, Sini JF, Delaunay D, Antoine MJ, Ducroz F, Garreau J, Griffiths R, Jones C, Lorin Y, Molle F, Péneau JP, Tétard Y, Violleau M (2000b) Experimental investigation of pollutant dispersion within a street in low wind conditions. In: 9th International Scientific Symposium Transport and Air Pollution, 5-8 Juin 2000, Avignon, France, Actes INRETS No 70, Ed. R. Joumard, vol 1, pp 95–102

Vachon G, Louka P, Rosant JM, Mestayer PG, Sini JF (2002) Measurements of traffic-induced turbulence within a street-canyon during Nantes '99 experiment. Water Air Soil Poll: Focus 2, No 5-6: 127-140

Wallenius L, Kukkonen J, Karppinen A, Walden J, Kartastenpää R, Koskentalo T, Aarnio P, Berkowicz R (2001) Evaluation of the OSPM model against the data measured during one year in Runeberg street, Helsinki. In: Cuvelier C et al.(eds) Seventh International Conference on Harmonisation within Atmospheric Dispersion Modelling for Regulatory Purposes. Joint Research Centre, European Commission, Ispra, Italy, pp 72–75


URL 4.1: http://medias.obs-mip.fr/escompte


# Chapter 5


Adams HS, Nieuwenhuijsen MJ, Colvile RN, McMullen MAJ, Khandelwal P (2001a) Fine particle (PM2.5) personal exposure levels in transport microenvironments, London, UK. Sci Total Environ 279(1-3): 29-44

Adams HS, Kenny LC, Nieuwenhuijsen MJ, Colvile R, Gussman R (2001b). Design and validation of a high flow personal sampler for $PM_{2.5}$. J Exposure Anal Env Med 11: 5-11

Adams HS, Nieuwenhuijsen MJ, Colvile RN (2001c) Determinants of fine particle ($PM_{2.5}$) personal exposure levels in transport microenvironments. London, UK. Atmos Environ 35: 4557-4566

Adams HS, Nieuwenhuijsen MJ, Colvile RN, Older MJ, Kendall M (2002) Assessment of road users' elemental carbon personal exposure levels. London, UK. Atmos Environ, Manuscript 21419, in press

Airborne Particles Experts Group (1999) Source apportionment of airborne partciulate matter in the United Kingdom. Department of Environment, Transport and Regions, UK

Areskoug H, Camner P, Dahlén SE, Låstbom L, Nyberg F, Pershagen G (ed), Sydbom, A. (2000) Particles in ambient air – a health risk assessment. Scand J Work Environ Health, 26, Supplement 1: 1–96

Benson PE (1984) CALINE 4 – A Dispersion Model for Predicting Air Pollutant Concentrations near Roadways, FHWA User Guide. Report No: FHWA/CA/TL-84/15, Trinity Consultants Inc., USA

Bowsher J (2000) Evaluation and development of numerical algorithms for multi-component aerosol modelling in London. PhD Thesis, Imperial College London

Bozó L (2000) Estimation of historical lead (Pb) deposition over Hungary. Időjárás 104: 161–172

Bozó L, Pinto J, Kelecsényi S (2001) Source-receptor relationships of fine aerosol particles in Budapest. In: Midgley P, Reuther M, Williams M (eds) Transport and Chemical Transformation in the Troposphere. Springer Verlag, pp 1115–1118

Gidhagen L, Johansson C, Ström J, Kristensson A Swietlicki E, Pirjola L (2002) Model simulation of ultrafine particles inside a Road tunnel. Atmos Environ, submitted. Also presented at the EUROTRAC-2 conference in Garmisch Partenkirchen, Germany, 2002

Harrison RM, Ping Shi J, Jones MR (1999) Continuous measurements of aerosol physical properties in the urban atmosphere. Atmos Environ 33: 1037–1047

Havasi Á, Bozó L, Zlatev Z (2001) Model simulation on the transboundary contribution to the atmospheric sulfur concentration and deposition in Hungary. Időjárás 105: 135–144

Hedberg E, Kristensson A, Ohlsson M, Johansson C, Johansson P-Å, Swietlicki E, Vesely V, Wideqvist U, Westerholm R (2002) Chemical and physical characterisation of emissions from birch wood combustion in a wood stove. Atmos Environ, in press

Johansson C, Hadenius A, Johansson PÅ, Jonson T (1999a) $NO_2$ and Particulate matter in Stockholm – Concentrations and population exposure. The Stockholm Study on Health effects of Air Pollution and their Economic Consequences. Swedish National Road Administration. Borlänge, Sweden. Available at http://www.slb.nu/lvf

Karppinen A, Härkönen J, Kukkonen J, Aarnio P, Koskentalo T (2002) Statistical model for assessing the portion of fine particulate matter transported regionally and long-range to urban air. Scand J Work Environ Health, 24 pp, in print

Kemp K, Palmgren F (2000) The Danish Air Quality Monitoring Programme. Annual Report for 1999. Department of Atmospheric Environment, 69 pp.- NERI Technical Report 357

Kerminen VM, Ojanen C, Pakkanen T, Hillamo R, Aurela M, Meriläinen J (2000) Low-molecular-weight dicarboxylic acids in an urban and rural atmosphere. Aerosol Sci 31: 349–362

Kleeman MJ, Schauer JJ, Cass GR (2000) Size and composition of fine particulate matter emitted from motor vehicles. Environ Sci Technol 34: 1132–1142

Kousa A, Kukkonen J, Karppinen A, Aarnio P, Koskentalo T (2002) A model for evaluating the population exposure to ambient air pollution in an urban area. Atmos Environ 36: 2109–2119

Kristensson A, Johansson C, Swietlicki E, Zhou J, Westerholm R, Wideqvist U, Vesely V (2000) Traffic source characterisation using factor analysis of the gas- and particle phase measured in a road tunnel. In: Midgley PM, and Reuther M (eds) Proc. EUROTRAC-2 Symposium 2002. Margraf Verlag, Weikersheim

Kristensson A, Johansson C, Swietlicki E, Zhou J, Westerholm R, Wideqvist U, Vesely V (2000) Traffic source characterisation using factor analysis of the gas- and particle phase measured in a road tunnel. In: Midgley PM, and Reuther M (eds) Proc. EUROTRAC-2 Symposium 2002. Margraf Verlag, Weikersheim

Kristensson A, Johansson C, Swietlicki E, Westerholm R, Wideqvist U, Vesely V, Zhou J, Papaspiropoulos G (2001) Emission of gases and particulate matter measured in a road tunnel. In: Proc. 3rd International Conference on Urban Air Quality — Measurement, modelling and management, 19–23 March, 2001, Loutraki, Greece

Kukkonen J, Härkönen J, Karppinen A, Pohjola M, Pietarila H, Koskentalo T (2001a) A semi-empirical model for urban $PM_{10}$ concentrations, and its evaluation against data from an urban measurement network. Atmos Environ 35: 4433–4442

Kukkonen J, Härkönen J, Walden J, Karppinen A, Lusa K (2001b) Evaluation of the dispersion model CAR-FMI against data from a measurement campaign near a major road. Atmos Environ 35-5: 949–960

Lai ACK, Nazaroff WW (2000) Modeling indoor particle deposition from turbulent flow onto smooth surfaces. J Aerosol Sci 31: 463

Laurikko JK (1998) On exhaust from petrol-fuelled passenger cars at low ambient temperatures. VTT Publications 348, Technical Research Centre of Finland, Espoo

McMurry PH (2000)A review of atmospheric aerosol measurements. Atmos Environ 34, 1959–1999

Morawska L, Johnson G, Ristovski ZD, Agranovski V (1999) Relation between particle mass and number for submicrometer airborne particles. Atmos Environ 33: 1983–1990

Pakkanen TA, Kerminen VM, Ojanen CH, Hillamo RE, Aarnio P, Koskentalo T (2000) Atmospheric black carbon in Helsinki. Atmos Environ 34: 1497–1506

Pakkanen TA, Kerminen VM, Korhonen CH, Hillamo RE, Aarnio P, Koskentalo T, Maenhaut W (2001a) Urban and rural ultrafine (PM$_{0.1}$) particles in the Helsinki area. Atmos Environ 35: 4593–4607

Pakkanen TA, Loukkola K, Korhonen CH, Aurela M, Mäkelä T, Hillamo RE, Aarnio P, Koskentalo T, Kousa A, Maenhaut W (2001b) Sources and chemical composition of atmospheric fine and coarse particles in the Helsinki area. Atmos Environ 35: 5381–5391

Pakkanen TA, Kerminen VM, Korhonen CH, Hillamo RE, Aarnio P, Koskentalo T, Maenhaut W (2001c) Use of atmospheric elemental size distributions in estimating aerosol sources in the Helsinki area. Atmos Environ 35: 5537–5551

Pakkanen TA, Hillamo RE (2002) Comparison of sampling artifacts and ion balances for a Berner low-pressure impactor and a virtual impactor. Boreal Environ Res 7: 129–140

Palmgren F, Wåhlin P, Berkowicz R, van Dingenen R (2001) Fine Particles from Traffic. In: Midgley P, Reuther M, Williams M (eds) Transport and Chemical Transformation in the Troposphere. Springer Verlag, pp 123–131

Pietarila H, Salmi T, Saari H, Pesonen R (2001) The preliminary assessment under the EC air quality directives in Finland SO$_2$, NO$_2$/NO$_x$, PM$_{10}$, lead. Report, Finnish Meteorological Institute

Pirjola L, Kulmala M (2000) Aerosol dynamical model MULTIMONO. Boreal Environ Res 5: 361–374

Pohjola M, Kousa A, Aarnio P, Koskentalo T, Kukkonen J, Härkönen J, Karppinen A (2000) Meteorological interpretation of measured urban PM$_{2.5}$ and PM$_{10}$ concentrations in the Helsinki Metropolitan Area. In: Longhurst JWS, Brebbia CA, Power H (eds) Air Pollution VIII. Wessex Institute of Technology Press, pp 689–698

Pohjola MA, Kousa A, Kukkonen J, Härkönen J, Karppinen A, Aarnio P, Koskentalo T (2002) The spatial and temporal variation of measured urban PM$_{10}$ and PM$_{2.5}$ concentrations in the Helsinki Metropolitan Area. Water Air Soil Poll: Focus 2, No 5-6: 189–201

Pohjola MA, Pirjola L, Kukkonen J, Kulmala M (2003) Modelling of the influence of aerosol processes for the dispersion of vehicular exhaust plumes in street environment. Atmos Environ 37, in print

Pope CA (2000) Review: Epidemiological basis for particulate air pollution health standards. Aerosol Sci Tech 32(1): 4–14

Quality of the Urban Air Review Group (1996) Airborne Particulate Matter in the United Kingdom. Published for the Department of Environment, Transport and Regions (DETR), UK

Scaperdas A, Colvile R (1999) Assessing the representativeness of monitoring data from an urban intersection site in Central London,UK. Atmos Environ 33: 661–674

San José R, Cortés J, Prieto JF, González RM (1998) Accurate ozone prognostic patterns for Madrid Area by using a high spatial and temporal Eulerian Photochemical Model. Environ Monit Assess 52: 203–212

Singles RJ, Carruthers DJ (1999) ADMS-Urban Air Quality Management System: Some examples of applications and validation in the UK. In: Proc. EUROTRAC Symposium 98. 821, WIT Press, Southampton

Sokhi RS, Fisher B, Lester A, McCrea I, Bualert S and Sootornsit N (1998) Modelling of air quality around roads. In: Proceedings from the 5th International Conference On Harmonisation with Atmospheric Dispersion Modelling for Regulatory Purposes. 18-21 May 1998, Greece, pp 492–497

Sokhi RS, Bualert S, Luhana L, Tremper A, Burton MA (2000a) Monitoring and modelling of $NO_2$ and fine particles for urban air quality management. SAMS, 37, pp 329–343

Sokhi RS, San Jose R, Moussiopoulos N, Berkowicz R(ed) (2000b) Urban Air Quality – Measurement, Modelling and Management. Kluwer Academic Publishers, pp 484

Sokhi RS, Bartzis JG (2002) Urban Air Qaulity – Recent Advances. A special issue of Water, Air and Soil Pollution: Focus 2, No5-6: pp 1-757

Tiitta P, Raunemaa T, Tissari J, Yli-Tuomi T, Leskinen A, Kukkonen J, Härkönen J, Karppinen A (2002) Measurements and modelling of $PM_{2.5}$ concentrations near a major road in Kuopio, Finland. Atmos Environ 36/25: 4057–4068

WHO (1999) Overview of the environment and health in Europe in the 1990's. WHO Regional Office of Europe, Executive Summary for the Third European Conference on Environment and Health, London (http://www.who.dk/london99/WelcomeE.htm)

WHO (2000) Methodology for assessment of environmental burden of disease: Report on the ISEE session on environmental burden of disease. Buffalo, 22 August 2000, WHO/SDE/WSH/00.7. http://www.who.int/environmental_information/Disburden/wsh007/Methodan1.htm

Wåhlin P, Palmgren F, van Dingenen R (2001a) Experimental studies of ultrafine particles in streets and the relationship to traffic. Atmos Environ 35: 63–69

Wåhlin P, Palmgren F, van Dingenen R, Raes F (2001b) Pronounced decrease of ambient particle number emissions from diesel traffic in Denmark after reduction of the sulphur content in diesel fuel. Atmos Environ 35/21: 3549-3552

URL 5.1: http://www.slb.nu.lvf
URL 5.2: http://www.airpollution.org.uk/dapple

# Chapter 6

Abart B, Sini JF, Mestayer PG (2000) First order turbulence modeling for the stratified atmospheric boundary layer. In: AMS 14th Symposium Boundary Layers and Turbulence. 7-11 August 2000, Aspen, Colorado

Almbauer RA (1995) A new finite volume discretisation for solving the Navier-stokes-equations. In: Numerical Methods in Laminar and Turbulent Flow. Pineridge Press, Swansea, UK, vol 9, pp 286–295

Almbauer R, Pucher K, Sturm PJ (1995) Air quality modelling for the city of Graz, Meteorology and Atmospheric Physics 57. Springer Verlag, Wien New York, pp 31–42

Anquetin S, Chollet JP, Copalle A, Mestayer P, Sini JF (1998) The urban atmosphere model SUBMESO. In: Borrell, PM, Borrell P (eds) Proc. EUROTRAC-2 Symposium '98. WITpress, Southampton, pp 750-757

Assimakopoulos V, Louka P, Moussiopoulos N, ApSimon H, Sahm P (2002) Evaluation of the microscale model MIMO with wind-tunnel measurements. In: Midgley PM, and Reuther M (eds) Proc. EUROTRAC-2 Symposium 2002. Margraf Verlag, Weikersheim 2002

Berkowicz R, Hertel O, Sorensen N, Michelsen J (1997) Modelling air pollution from traffic in urban areas. In: Perkins R, Belcher S (eds) Flow and Dispersion through Groups of Obstacles. Clarendon Press, Oxford, pp 121–141

Berkowicz R, Ketzel M, Vachon G, Louka P, Rosant JM, Mestayer PG, Sini JF (2002) Examination of traffic pollution distribution in a street using the Nantes'99 experimental data and comparison with model results. Water Air Soil Poll: Focus 2, No 5-6: 311-324

Borrego C, Martins JM, Tome M, Carvalho A, Barros N, Pinto C (2001) Wind tunnel validation of VADIS, a numerical model simulating flow and dispersion around building sets. In: Midgley PM, Reuther M, Williams M (eds) Proc. EUROTRAC-2 Symposium 2000. Springer-Verlag, Berlin

Carruthers DJ, Edmunds HA, Bennet M, Woods PT, Milton MJT, Robinson R, Underwood BY, Franklin CJ (1997) Validation of the ADMS dispersion model and assessment of its performance relative to R-91 and ISC using archived LIDAR data. Int J Environ Poll 8, Nos 3-6, 264–278

Carruthers DJ, Dixon P, McHugh CA, Nixon SG, Oates W (1999a) Determination of compliance with uk and eu air quality objectives from high resolution pollutant concentration maps calculated using ADMS-Urban. In: Proc. 6th Conference on Harmonisation within dispersion modelling for regulatory purposes, October 1999, Rouen, France, to be published in Int J Environ Pollut

Carruthers DJ, McHugh CA, Nixon SG (1999b) Use of ADMS-Urban to calculate high resolution air quality maps in urban areas. CERC, 3 King's Parade, Cambridge, CB2 1SJ, UK

DePaul FT, Sheih CM (1986) Measurements of wind velocities in a street canyon. Atmos Environ 20: 455–459

Dierer S (1997) Evaluierung des mesoskaligen Transport- und Strömungsmodells METRAS. Diplomarbeit, Fachbereich Geowissenschaften, Universität Hamburg, pp 120

Dupont S, Guilloteau E, Mestayer PG (2000) Energy balance and surface temperatures of urban quarters. In: Proc. AMS 3rd Symposium on Urban Environment. 14-18 August 2000, Davis, California, pp 149–150

Dupont S, Calmet I, Mestayer PG (2002) Urban canopy modeling influence on urban boundary layer simulation, In: Proceedings of the AMS 4th symposium on Urban Climatology. Norfolk, VA

Ehrhard J, Ernst G, Goetting J, Khatib A, Kunz R, Moussiopoulos N, Winkler C (1999) The microscale model MIMO: Development and Assessment

Ehrhard J, Ernst G, Khatib IA, Kunz R, Moussiopoulos N, Winkler C (2000) The Microscale Model MIMO: development and assessment. J Wind Eng Ind Aerodyn 85(2): 163–176

Galmarini S, Thunis P (2000) Mesoscale model interComparison: Results Analysis. http://rem.jrc.cec.eu.int/~galmarin/mesocom/results/index.html

Goetting J, Winkler C, Rau M, Moussiopoulos N, Ernst G (1997) Dispersion of a passive pollutant in the vicinity of a U-shaped building. Int J Environ Pollut 8, Nos 3-6: 718–726

Gronskei KE, Walker SE, Gram F (1992) Evaluation of a model for hourly spatial concentrations distributions. Atmos Environ 27B: 105-120

Härkönen J, Valkonen E, Kukkonen J, Rantakrans E, Jalkanen L, Lahtinen K (1995) An operational dispersion model for predicting pollution from a road. Int J Environ Pollut 5, Nos 4-6: 602–610

Jicha M, Katolicky J, Pospisil J (2000) Dispersion of Pollutants in street canyon under traffic induced flow and turbulence. Int J Environ Monitor Assess 65: 343–351

Karppinen A, Kukkonen J, Elolähde T, Konttinen M, Koskentalo T (2000a) A modelling system for predicting urban air pollution, Comparison of model predictions with the data of an urban measurement network. Atmos Environ 34-22 pp 3735–3743

Karppinen A, Kukkonen J, Elolähde T, Konttinen M, Koskentalo T, Rantakrans E (2000b) A modelling system for predicting urban air pollution, Model description and applications in the Helsinki metropolitan area. Atmos Environ 34-22, pp 3723–3733

Kastner-Klein P, Fedorovich E, Sini JF, Mestayer PG (2000) Experimental and numerical verification of similarity concept for dispersion of car exhaust gases in urban street canyons. Int J Environ Monitor Assess 65: 353–361

Kastner-Klein P, Berkowicz R, Fedorovich E (2001) Evaluation of Scaling Concepts for Traffic-Produced Turbulence Based on Laboratory and Full-Scale Concentration Measurements in Street Canyons. In: Proc. 3rd International Conference on Urban Air Quality. 19-23 March 2001, Loutraki

Kessler E (1969) On the distribution and continuity of water substance in the atmospheric circulations. Meteor Monogr 10 (32)

Ketzel M, Louka P, Sahm P, Guilloteau E, Sini JF, Moussiopoulos N (2002) Intercomparison of Numerical Urban Dispersion Models - Part II: Street Canyon in Hannover. Germany. Water Air Soil Pol.: Focus 2, No 5-6: 603-613

Kousa A, Kukkonen J, Karppinen A, Aarnio P, Koskentalo T (2001) Statistical and diagnostic evaluation of a new-generation urban dispersion modelling system against an extensive dataset in the Helsinki Area. Atmos Environ, 35/27: 4617–4628

Kovar-Panskus A, Louka P, Mestayer PG, Savory E, Sini JF, Toy N (2001) Influence of Geometry on the Flow and Turbulence Characteristics Within Urban Street Canyons - Intercomparison of Wind Tunnel Experiments and Numerical Simulations. Water Air Soil Poll: Focus 2, No 5-6: 365-380

Kukkonen J, Härkönen J, Walden J, Karppinen A, Lusa K (2001a) Evaluation of the dispersion model CAR-FMI against data from a measurement campaign near a major road. Atmos Environ 35: 949–960

Kukkonen J, Kousa A, Karppinen A, Aarnio P, Koskentalo T (2001b) A Model of the Exposure of Population to Ambient Air Pollution, and Numerical Results in Helsinki. Proc of the 3rd International Conference on Urban Air Quality, Measurement, Modelling and Management. 19-23 March 2001, Loutraki, Greece

Kunz R, Moussiopoulos N (1995) Simulation of the wind field in Athens using refined boundary conditions. Atmos Environ 29: 3575–3591

Kunz R, Khatib I, Moussiopoulos N (1998) Coupling of Mesoscale and Microscale models-An approach to simulate scale interaction

Larssen S, Gronskei KE, Gram F, Hagen LO, Walker SE (1994) Verification of urban scale time dependent dispersion model with subgrid elements in Oslo, Norway. In: Air Pollution Modelling and ist Appl. X. Plenum Press, New York

de Leeuw F, Moussiopoulos N, Sahm P, Bartonova A (2002) Urban air quality in larger conurbations in the European Union. Environ Modell Softw 16: 399–414

Lenz CJ, Müller F, Schlünzen KH (2000) The sensitivity of mesoscale chemistry transport model results to boundary values. Environ Monit Assess 65: 287–298

Louka P, Vachon G, Sini JF, Mestayer PG, Rosant JM (2002) Thermal effects on the airflow in a street canyon - Nantes '99 experimental results and model simulation. Water Air Soil Poll: Focus 2, No 5-6: 351-364

Lüpkes C, Schlünzen KH (1996) Modelling the Arctic convective boundary-layer with different turbulence parameterizations. Boundary-Layer Meteorol 79: 107–130

Martilli A, Clappier A, Rotach MW (2000) The urban atmospheric boundary layer: A modelling study. In: 16th IMACS World Congress 2000, Session on Urban Air pollution Modeling. 21-25 August, 2000, Lausanne, Switzerland

Martins JM, Borrego C (1998) Describing the dispersion of pollutants near buildings under low wind-speeds: real scale and numerical results. Proc. Envirosoft 98, WIT, Las Vegas

Mensink C, De Ridder K, Lewyckyj N, Delobbe L, Van Haver Ph, Janssen L (2001) Air quality modelling in Urban Regions using an Optimal Resolution Approach (AURORA). In: Proc. 7th International Conference on Harmonisation within Atmospheric Dispersion Modelling for Regulatory Purposes. May 2001,Belgirate, European Commission, Joint Research Centre, Ispra, pp 459–463

Mestayer PG (1996) Development of the French Communal Model SUBMESO for Simulating Dynamics, Physics and Photochemistry of the Urban Atmosphere. In: Borrell PM, Borrel P, Cvitas T, Kelly K, Seiler W (eds) Proc. EUROTRAC-2 Symposium '96. Computational Mechanics Public., Southhampton, pp 539–544

Moussiopoulos N (1987) An efficient scheme to calculate radiative transfer in mesoscale models. Environ Softw 2(4): 172–191

Moussiopoulos N (1989) Mathematische Modellierung mesoskaliger Ausbreitung in der Atmosphaere. Fortschr.-Ber., VDI, Reihe 15, 64: 307

Moussiopoulos N (1995) The EUMAC Zooming Model, a tool for local-to-regional air quality studies. Meteorol Atmos Phys 57: 115–133

Moussiopoulos N, Papagrigoriou S (eds) (1997) Athens 2004 Air Quality. In: Proc. International Scientific Workshop "Athens 2004 Air Quality Study". 18-19 February 1997, Athens, 183 pp Available also as a CD-ROM from http://www.envirocomp.org/

Moussiopoulos N, Sahm P (1998) The OFIS model: An efficient tool for assessing ozone exposure and evaluating air pollution abatement strategies. In: Proc. 5th Conference on Harmonization within Atmospheric Dispersion Modelling for Regulatory Purposes, May 1998, Rhodos

Moussiopoulos N, Sahm P (2000) The OFIS model: an efficient tool for assessing ozone exposure and evaluating air pollution abatement strategies. Int J Environ Pollut 14: 597–606

Moussiopoulos N, Sahm, Kessler Ch (1995) Numerical simulation of photochemical smog formation in Athens, Greece-a case study. Atmos Environ 29: 3619–3632

Moussiopoulos N, Sahm P, Kunz R, Vögele T, Schneider Ch, Kessler Ch (1997) High resolution simulations of the wind flow and the ozone formation during the Heilbronn ozone Experiment. Atmos Environ 31: 3177–3186

Müller F, Schlünzen KH, Schatzmann M (2000) Test of numerical solvers for chemical reaction mechanisms in 3D air quality models. Environ Modell Soft 15: 639–646

Öttl D (2000) Weiterentwicklung, Validierung und Anwendung eines mesoskaligen Modells. Dissertation, Karl Franzens Universität Graz, pp 155

Öttl D, Almbauer RA, Sturm PJ (2001) A new method to estimate diffusion in stable, low wind conditions. J Applied Meteorol 40: 259–268

Öttl D, Kukkonen J, Almbauer RA, Sturm PJ, Pohjola M, Härkönen J (2001) Evaluation of a Gaussian and a Lagrangian model against a roadside dataset, with focus on low wind speed conditions. Atmos Environ 35: 2123–2132

Panskus H (2000) Konzept zur Evaluation hindernisauflösender mikroskaliger Modelle und seine Anwendung auf das Modell MITRAS. Fortschritt-Berichte VDI, VDI-Verlag, Düsseldorf, 7-389, pp 97

Panskus H, Schlünzen KH (1997) Standards for writing and documentation Fortran 90 code for the MITRAS/METRAS model system. MITRAS Technical Report Nr. 1. http://www.mi.uni-hamburg.de/technische_meteorologie/Meso/mitras/mitras.html

Pénelon T, Calmet I, Mestayer P (2002) Influence of a Small-Scale Topography on the Dynamics of Atmospheric Boundary Layer Flows. In: 15th Symposium on Boundary Layers and Turbulence, 15-19 July 2002, Wageningen, The Netherlands, submitted

Qin Y, Kot SC (1993) Dispersion of vehicular emission in street canyons, Guangzhou city, South china (P.R.C.). Atmos Environ 27B: 283–291

Sahm P, Moussiopoulos N (1995) MUSE - a multilayer dispersion model for reactive pollutants, in Air Pollution III. In: Power H, Moussiopoulos N, Brebbia CA (eds) Computational Mechanics Publications, Southampton, 1: 359–368

Sahm P, Moussiopoulos N (1996) MUSE - A new three layer photochemical dispersion model. In: Borrell PM, Borrel P, Cvitas T, Kelly K, Seiler W (eds) Proc. EUROTRAC Symposium '96. Computer Mechanics Publications, Southampton, pp 553–557

Sahm P, Moussiopoulos N (1998) A new approach for assessing ozone exposure and for evaluating control strategies at the urban scale. In: Proc. of the Air Pollution 98, September 1998, Genoa

Sahm P, Moussiopoulos N (1999) The OFIS model: A new approach in urban scale photochemical modeling. EUROTRAC Newsletter 21, pp 22–28

Sahm P, Kirchner F, Moussiopoulos N (1997) Development and Validation of the Multilayer Model MUSE - The Impact of the Chemical Reaction Mechanism on Air Quality Predictions. In: Proc. of the 22nd NATO/CCMS International Technical Meeting on Air Pollution Modelling and its Application. June 2-6, 1997, Clermont-Ferrand

Sahm P, Moussiopoulos N, Theodoridis G, Assimakopoulos V, Berkowicz R (2000) Chapter 1.5 - The Influence of Fast Chemistry on the Composition of $NO_x$ in the Emission Input to Atmospheric Dispersion Models in TRAPOS final report

Sahm P, Louka P, Ketzel M, Guilloteau E, Sini JF (2002) Intercomparison of Numerical Urban Dispersion Models - Part I: Street Canyon and Single Building Configurations. Water Air Soil Poll: Focus 2, No 5-6: 587-601

von Salzen K, Claussen M, Schlünzen KH (1996) Application of the concept of blending height to the calculation of surface fluxes in a mesoscale model. Meteorol Zeitschrift, N.F. 5, pp 60–66

Scaperdas A, Colvile RN, Robins AG (2000) Understanding flow patterns at street canyon intersections using wind tunnel and CFD simulations. In: Borrell PM, Borrell P (eds) Proc. EUROTRAC-2 Symposium 1998. WITpress, Boston Southampton, pp 796

Scaperdas A, Robins AG, Colvile RN (2001) Microscale dispersion modelling in cities: the effect of street-building arrangements. Environ Monit Assess, submitted

Schlünzen KH (1990) Numerical studies on the inland penetration of sea breeze fronts at a coastline with tidally flooded mudflats. Beitr Phys Atmosph 63: 243–256

Schlünzen KH (1994) Mesoscale modelling in complex terrain - an overview on the German nonhydrostatic models. Beitr Phys Atmosph 67: 243–253

Schlünzen KH (1996) Validierung hochauflösender Regionalmodelle. Ber. aus dem Zentrum f. Meeres- und Klimaforschung, Meteorologisches Institut, Universität Hamburg, A23, pp 184

Schlünzen KH (1997) On the validation of high-resolution atmospheric mesoscale models. J Wind Eng Ind Aerodyn 67 & 68: 479–492

Schlünzen KH (2001) Model inventory for EUROTRAC-2 subproject SATURN. In: Midgley PM, Reuther M, Williams M (eds) Proc. EUROTRAC-2 Symposium 2000, Springer Verlag, Berlin

Schlünzen KH (2002) SATURN model overview
    http://www.mi.uni-hamburg.de/technische_meteorologie/Meso/saturn/overview.html

Schlünzen KH (2002) Simulation of transport and chemical transformations in the atmospheric boundary layer - review on the past 20 years developments in science and practice. Meteorol Z 11, in print

Schlünzen KH, Hinneburg D, Knoth O, Lambrecht M, Leitl B, Lopez S, Lüpkes C, Panskus H, Renner E, Schatzmann M, Schoenemeyer T, Trepte S, Wolke R (2002) Flow and transport in the obstacle layer - First results of the microscale model MITRAS. J Atmos Chem, in print

Silibello C, Calori G, Brusasca G, Catenacci G, Finzi G (1998) Application of a photochemical grid model to Milan metropolitan area. Atmos Environ 32: 2025–2038

Sini JF, Anquetin S, Mestayer PG (1996). Pollutant dispersion and thermal effects in urban street canyons. Atmos Environ 30: 2659–2677

Sokhi RS, Fisher B, Lester A, McCrea I, Bualert S, Sootornstit N (1998) Modelling of air quality around roads. In: Proc. 5th International Conference of Harmonisation within Atmospheric Dispersion Modelling for Regulatory Purposes. 18-21 May 1998, Rhodes, pp 492–497

Stephens GL (1978) Radiation profiles in extended water clouds. II: Parameterization schemes, J Atmos Sci 35: 2123–2132

Vachon G, Louka P, Rosant JM, Mestayer PG, Sini JF (2002) Measurements of traffic-induced turbulence within a street-canyon during Nantes '99 experiment. Water Air Soil Poll: Focus 2, No 5-6: 127-140

Yamartino RJ, Scire JS, Carmichael GR, Chang YS (1992) The CALGRID mesoscale photochemical grid model-I: Model formulation. Atmos Environ 26A: 1493–1512

Yamartino RJ (1993) Nonnegative, conserved scalar transport using grid-cell-centered, spectrally constrained Blackman cubics for applications on a variable-thickness mesh. Mon Weather Rev 121: 753–763


URL 6.1: http://www.dmu.dk/AtmosphericEnvironment/TRAPOS
URL 6.2: http://www.mi.uni-hamburg.de/cedval
URL 6.3: http://reports.eea.eu.int:80/Topic_report_No_032001

# Chapter 7

Borrego C, Barros N, Miranda AI, Carvalho AC, Valinhas MJ (1998) Validation of two photochemical numerical systems under complex mesoscale circulations. Proc. 23rd NATO/CCMS Int. Tech. Meeting on Air Pollution Modelling and its Application, Bulgaria

Borrell P (1998) Quality Assurance and Quality Control in EUROTRAC-2. A statement by the Scientific Steering Committee (SSC). IFU Garmish-Paterkirchen

Britter RE (1994) The Evaluation of technical models used for major-accident hazard installation. – Report EUR 14774 EN, Brussels

Eichhorn J (1989) Entwicklung und Anwendung eines dreidimensionalen mikroskaligen Stadtklima-Modells. Dissertation, Universität Mainz, Germany

Fox DG (1984) Uncertainty in Air Quality Modelling. B Am Meteorol Soc 65, No 1: 27–37

Galmarini S, Bianconi R, Bellasio R, Graziani G (2001) Forecasting consequences of accidental releases from ensemble dispersion modelling. J Environ Radioactiv 57: 203–219

Girardi F, Graziani G, van Veltzen D, Galmarini S, Mosca S, Bianconi R, Bellasio R, Klug W (eds) (1998) The ETEX project. EUR Report 181-43 EN. Office for official publications of the European Communities, Luxembourg, 108 pp

Graziani G, Mosca S, Klug W (1998a) Real-time long-range dispersion model evaluation of ETEX first release. EUR 17754/EN. Luxembourg: Office for Official Publications of the European Commission

Graziani G, Klug W, Galmarini S, Grippa G (1998b) Real-time long-range dispersion model evaluation of ETEX second release. EUR 17755/EN. Luxembourg: Office for Official Publications of the European Commission

Hall RC (ed) (1997) Evaluation of modelling uncertainty - CFD modelling of near-field atmospheric dispersion. EU Project EV5V-CT94-0531, Final Report. WS Atkins Consultants Ltd., Woodcote Grove, Ashley Road, Epsm, Surrey KT18 5BW, UK

Hanna SR (1994) Mesoscale meteorology model evaluation techniques with emphasis on needs of air quality models. In: Peilke RA, Pearce Sr and RP (eds) Mesoscale Modeling of the Atmosphere, American Meteorological Society, Boston, pp 47–58

Hanna SR, Yang R (2001) Evaluation of mesoscale models' simulations of near-surface winds, temperatures and mixing depths. J Applied Meteorol 40: 1095–1104

ISO 14050:1998 – Environmental Management – Vocabulary

Ketzel M, Berkowicz R, Lohmeyer A (1999) Dispersion of traffic emissions in street canyons - Comparison of European numerical models with each other as well as with results from wind tunnel and field measurements. In: Proc. 2nd International Conference on Urban Air Quality, March 3-5, Madrid

Ketzel M, Louka P, Sahm P, Guilloteau E, Sini JF, Moussiopoulos N (2002) Intercomparison of Numerical Urban Dispersion Models - Part II: Street Canyon in Hannover. Germany. Water Air Soil Pol.: Focus 2, No 5-6: 603-613

Klug W, Graziani G, Grippa G, Pierce D, Tassone C (eds) (1992) Evaluation of long range atmospheric transport models using environmental radioactivity data from the Chernobyl accident. EUR Report 14147 EN, Office for official publications of the European Communities, Luxembourg, 366 pp

Leitl B (2000) Validation Data for Microscale Dispersion Modelling. EUROTRAC Newsletter 22, pp 28–32

Martinez JR, Javitz HS, Ruff RE, Valdes A, Nitz KC Dabberdt WF (1981) Methodology for Evaluation Highway Air Pollution Dispersion Models. Transportation Research Board, USA

MEG- Model Evaluation Group, Report on the second Open Meeting. By Cole ST and PJWicks, France (1994)

Mosca S, Graziani G, Klug W, Bellasio R, Bianconi R (1998a) ATMES-II - Evaluation of long-range dispersion models using first ETEX release data. EUR 17756/EN. Luxembourg: Office for Official Publications of the European Commission

Mosca S, Graziani G, Klug W, Bellasio R, Bianconi R (1998b) A statistical methodology for the evaluation of long-range dispersion models: an application to the ETEX exercise. Atmos Environ 32, No 24: 4307–4324

Moussiopoulos N, Berge E, Bøhler T, de Leeuw F, Grønskei KE, Mylona S, Tombrou M (1996) Ambient Air Quality, Pollutant Dispersion and Transport Models. European Topic Centre on Air Quality, Topic Report 19, 1996, European Environment Agency, EU Publications, Copenhagen, 94 pp

Moussiopoulos N, de Leeuw F, Karatzas K, Bassoukos A (2000) The air quality model documentation system of the European Environment Agency. Int J Environ Pollut 14: 10–17

Moussiopoulos N, Sahm P, Münchow S, Tonnesen D, de Leeuw F, Tarrasón L (2001) Uncertainty analysis of modeling studies included in air quality assessments. Int J Environ Pollut, in press

NLÖ (1994) Lufthygienisches Überwachungssystem Niedersachsen - Luftschadstoffe in Straßenschluchten. Bericht, Niedersächsichses Landesamt für Ökologie, Göttinger Str. 14, 30449 Hannover, ISSN 0945 4187

NLÖ (1995) Lufthygienisches Überwachungssystem Niedersachsen - Standortbeschreibung der LÜN Stationen. Bericht, Niedersächsichses Landesamt für Ökologie, Göttinger Str. 14, 30449 Hannover, ISSN 0945 4187 (in German)

Pielke RA Sr (2002) Mesoscale meteorological modelling. 2nd edn. Academic Press, New York, 200, pp 676

Schädler G, Bächlin W, Lohmeyer A, van Wees T (1996) Vergleich und Bewertung derzeit verfügbarer mikroskaliger Strömungs- und Ausbreitungsmodelle. In: Berichte Umweltforschung Baden-Württemberg, PEF 293001, (FZKA-PEF 138) (in German)

Schatzmann M, Leitl B (2002) Validation and Application of obstacle resolving urban dispersion models. Atmos Environ 36: 4811–4821

Schlünzen KH (1997) On the validation of high-resolution atmospheric mesoscale models. J Wind Eng Ind Aerodyn 67-68: 479–492

286

Thunis P, Galmarini S, Martilli A, Clappier A, Andronopoulos S, Bartzis J, Vlachogianni M, deRidder K, Moussiopoulos N, Sahm P, Almbauer R, Sturm P, Oettl D, Dierer S and Schluezen H (2002) MESOCOM: An inter-comparison exercise of mesoscale flow models applied to an ideal case simulation. To appear in Atmos Environ

US EPA – Guideline for Regulatory Application of the Urban Airshed Model. EPA-450/4-91-013 (1991)

VDI-Richtlinie 3783, Blatt 9, 2002: Umweltmeteorologie. Prognostische mikroskalige Windfeldmodelle. Evaluierung für Gebäude- und Hindernisumströmung. Vor-Entwurf vom 15.1.2002 (in German)

URL 7.1 http://www.gsf.de/eurotrac/g-qa-qc.htm
URL 7.2: http://rem.jrc.cec.eu.int/mesocom/
URL 7.3 http://rem.jrc.cec.eu.int/escompte_int
URL 7.4: http://www.mi.uni-hamburg.de/cedval

# Chapter 8

Barros N, Borrego C (1995) Influence of coastal breezes on the photochemical production over the Lisbon region. In: Moussiopoulos N, Power H, Brebbia CA (eds) Air Pollution III, Vol 3: Urban Pollution. Computational Mechanics Publ., pp 67–74

Bhugwant C, Cachier H, Bremaud P, Roumeau S, Leveau J (2000) Chemical effect of carbonaceous aerosols on the diurnal cycle of MBL ozone at a tropical site: measurements and simulations. Tellus B 52: 1232–1248

Borrego C, Barros N, Coutinho M (1994): Application of two mesometeorological models to the Lisbon region: Preliminary conclusions. EUROTRAC Symposium '94, 11-15 April, Garmisch-Partenkirchen, Germany, pp 885

Borrego C, Barros N, Miranda AI, Carvalho AC, Valinhas MJ (1998) Validation of two photochemical numerical systems under complex mesoscale circulations. 23rd Int. Tech. Meeting of NATO/CCMS on Air Pollution Modelling and its Application, September 28-October 2, Varna, Bulgaria, pp 411–418

Borrego C, Barros N, Tchepel O (2000) An assessment of modelling ozone control abatement strategies in Portugal: the Lisbon urban area. Int. Tech. Meeting of NATO/CCMS on Air Pollution Modelling and its Application. 15-19 May, Boulder, Colorado, USA, pp 439–446

Borghi S, Giuliacci M (1980) Circulation features driven by diurnal heating in the lower atmospheric layers in the Po Valley. Il Nuovo Cimento, vol 3C, No 1

Castro T, Madronich S, Rivale S, Muhlia A, Mar B (2001) The influence of aerosols on photochemical smog in Mexico City. Atmos Environ 35: 1765–1772

Decanini E, Volta M (2002) Flexible modelling system for air pollution simulation and control in Northern Italy. In: Midgley PM and Reuther M (eds) Proc. EUROTRAC-2 Symposium 2002, Margraf Verlag, Weikersheim

Dickerson RR, Kondragunta S, Stenchikov G, Civerolo KL, Doddridge BG, Holben BN (1997) The impact of aerosols on solar ultraviolet radiation and photochemical smog. Science 278: 827–830

Dosio A, Emeis S, Graziani G, Junkermann W, Levy A (2001) Assessing the meteorological conditions in a deep Alpine valley system by a measuring campaign and simulation with two models during a summer smog episode. Atmos Environ 35: 5441–5454

EMEP (1999) Transboundary Photo-oxidant air pollution in Europe. EMEP/MSC-W Status report 2/99. DNMI. Norway

ENEL/Prod ULP (1998) Riconfigurazione della RRQA e Adempimenti Ambientali Concordati con la Provincia di Savona ai Sensi del Decreto M.I.C.A. del 23.06.93. - Caratterizzazione Meteorologica e della Qualità dell'Aria del Periodo Estivo, Relazione Tecnica 212VL11922

Environment and Systems (1999) Photochemical pollution in the Basque Country Autonomous Community. Servicio Central de Publicaciones del Gobierno Vasco. Departamento de Ordenacion del Territorio, Vivienda yMedio Ambiente. Vitoria-Gasteiz, Spain

Finardi S, Tinarelli G, Nanni A, Brusasca G, Carboni G (1999) Evaluation of a 3D Flow and Pollutant Dispersion Modelling System to Estimate Climatological Ground Level Concentrations in Complex Coastal Sites. In: Proc. of the 6th International Conference on Harmonization within Atmospheric Dispersion Modelling for Regulatory Purposes. October 11-14, Rouen, France

Finzi G, Volta M (2000) Real time urban ozone alarm system including neuro-fuzzy networks and grey box models. In: Proc. International ICSC/IFAC Symposium on NEURAL COMPUTATION-NC2000, ICSC Academic Press

Finzi G, Silibello C, Volta M (2000) Evaluation of urban pollution abatement strategies by a photochemical dispersion model. Int J Environ Pollut 14: 1–6

Finzi G, Volta M, Carnevale C, Ragni C (2002) POEMPM: an emission model for size and chemical description of particulate matter. In: Midgley PM, Reuther M (eds) Proc. EUROTRAC-2 Symposium 2002. Margraf Verlag, Weikersheim

Gabusi V, Volta M, Veraldi S, Veronesi V (2001) Road Traffic Impact on Photochemical Pollution: Brescia Metropolitan Area Case. In: Midgley PM, Reuther M, Williams M (eds) Proc. EUROTRAC-2 Symposium 2000, Springer Verlag Berlin

Gabusi V, Pertot C, Pirovano G, Volta M (2002) First results of a long-term simulation of photochemical smog in Northern Italy. In: P.M. Midgley, M. Reuther (eds) Proc EUROTRAC-2 Symposium 2002, Margraf Verlag, Weikersheim

Gaffney JS, Marley NA, Drayton PJ, Doskey PV, Kotamarthi VR, Cunningham MM, Baird JC, Dintaman J, Hart HL (2002) Field observations of regional and urban impacts on $NO_2$, ozone, UVB, and nitrate radical production rates in the Phoenix air basin. Atmos Environ 36: 825–833

Gangoiti G, Millán MM, Salvador R, Mantilla E (2001) Long-range transport and recirculation of pollutants in the western Mediterranean during the project Regional Cycles of Air Pollution in the West-Central Mediterranean Area. Atmos Environ 35: 6267–6276

Gangoiti G, Alonso L, Navazo M, Albizuri A, Perez-Landa G, Matabuena M, Valdenebro V, Maruri M, García JA, Millán MM (2002) Regional transport of pollutants over the Bayof Biscay: analysis of an ozone episode under a blocking anticyclone in west-central Europe. Atmos Environ 36: 1349–1361

He S, Carmichael GR (1999) Sensitivity of photolysis rates and ozone production in the troposphere to aerosol properties. J Geophys Res 104: 26307–26324

Hogrefe C, Rao ST, Kasibhatla P, Hao W, Sistla G, Mathur R, McHenry J (2001) Evaluating the performances of regional-scale photochemical modeling systems: Part II Ozone predictions. Atmos Environ 35: 4175–2188

IPCC 95 (1996) Climate Change 1995: The Science of Climate Change. Houghton JT, Meira Filho LG, Callander BA, Harris N, Kattenberg A, Maskell K (eds), Cambridge University Press

Kalabokas PD, Bartzis JG (1998) Photochemical air pollution characteristics at the station of the NCSR-Demokritos, during the MEDCAPHOT-Trace campaign in Athens, Greece (20 August – 20 September 1994). Atmos Environ 32: 2123–2139

Kallos G, Kassomenos P, Pielke R (1993) Synoptic and mesoscale weather condotions during air pollution episodes in Athens, Greece, Boundary-Layer Meteorol 62: 163–184

Kasibhatla P, Chameides WL (2000) Seasonal modeling of regional ozone pollution in the Eastern United States. Geophys Res Lett 27: 1415–1418

Kassomenos P, Flocas HA, Lycoudis S, Petrakis M (1998a) Analysis of mesoscale patterns in relation to synoptic conditions over an urban Mediterranean basin. Theor Appl Climatol 59: 215-219

Kassomenos PA, Flocas HA, Lykoudis S, Skouloudis A (1998b) Spatial and temporal characteristics of the relationship between air quality status and mesoscale circulation over an urban Mediterranean basin. Sci Total Environ 217: 37–57

Kondratyev Kya, Varotsos CA (2000) Investigation tropospheric ozone in Europe. Ecological Chem 9(1-2): 3–9

Kruger BC, Kroger H, Wotawa G, Kromp-Kolb H (2000) Lagrangian Photochemical Model Calculations for the Milan Area. In: Midgley PM, Reuther M, Williams M (eds) Proc. EUROTRAC-2 Symposium 2000, Springer Verlag, Berlin

Lawson DR (1990) The southern California air quality study. J Air Waste Manage Assoc 40: 156-165

Lee DS, Holland MR, Falla N (1996) The potential impact of ozone on materials in the U.K. Atmos Environ 30: 1053–1065

de Leeuw F, Bogman F (2001) Air pollution by ozone in Europe in summer 2001 - Overview of exceedances of EC ozone threshold values during the summer season April–August 2001. EEA Topic Report No 13/2001. European Environmental Agency, Copenhagen

de Leeuw F, de Paus TA (2001) Exceedance of EC ozone threshold values in Europe in 1997. Water Air Soil Poll 128: 255–281

de Leeuw F, Sluyter R, van Zantvoort E (1998) Exceedance of Ozone Threshold Values in 1996 and Summer 1997. European Topic Centre on Air Quality of the European Environment Agency. Topic Report 97/10. Office for Official Publications of the European Communities, Luxembourg

Milford J, Gao D, Sillman S, Blossery P, Russel AG (1994) Total reactive nitrogen ($NO_y$) as an indicator for sensitivity of ozone to $NO_x$ and hydrocarbons. J Geophys Res 99: 3533–3542

Millán MM, Sanz MJ, Salvador R, Mantilla E (2002) Atmospheric dynamics and ozone cycles related to nitrogen deposition in the western Mediterranean. Environ Pollut 118: 167–186

Monteiro A, Lopes M, Borrego C, Miranda AI (2002) Contribution of air pollution to the managment of carbon cycle on a Portuguese coastal region. In: Brebbia CA (ed) Coastal Environment 2002. 14-19 September, Rhodes, Greece, 2002, Environmental Coastal Regions VI, WITpress Southampton, pp 395–404

Moussiopoulos N, Sahm P, Tourlou PM, Friedrich R, Wickert B, Reis S, Simpson D (1998) Technical Expertise in the Context of the Commission's Communication on an Ozone Strategy. Final report to European Commission DGXI, pp 36

Moussiopoulos N, Papalexiou S, Lammel G, Arvanitis A (2000) Simulation of nitrous acid formation taking into account heterogeneous pathways: Application to the Milan metropolitan area. Environ Modell Softw 15: 629–637

Neftel A, Clappier A, Favaro G, Martilli A (2000) Simulation of the Ozone Formation in the Northern Part of the Po Valley with the TVM-CTM. In: EUROTRAC-2 Symposium 2000. March 27-31, 2000, Garmish-Partenkirchen

Neftel A, Spirig C, Kleinman L, Sillman S (2002) The VOC/NOx sensitivity of the ozone production in the troposphere - an important issue?! LOOP Gerzensee Background document, pp 1–51

Prevot ASH, Staehelin J, Kok GL, Schillawski RD, Neininger B, Staffelbach T, Neftel A, Wernli H, Dommen J (1997) The Milan photooxidant plume. J Geophys Res 102: 23375-23388

Pujadas M, Plaza J, Terés J, Artíñano B, Millán M (2000) Passive remote sensing of nitrogen dioxide as a tool for tracking air pollution in urban areas: the Madrid urban plume, a case of study. Atmos Environ 34: 3041–3056

de Reus M, Dentener F, Thomas A, Borrmann S, Strom J, Lelieveld J (2000) Airborne observations of dust aerosol over the North Atlantic Ocean during ACE 2: Indications for heterogeneous ozone destruction. J Geophys Res 105: 15263–15275

Sahm P, Moussiopoulos N, Theodoridis G, Assimakopoulos V, Berkowicz R (2000) Chapter 1.5 - The Influence of Fast Chemistry on the Composition of NOx in the Emission Input to Atmospheric Dispersion Models in TRAPOS final report

San José R, Salas I (2000) A sensitivity analysis study of the impact of traffic (city and surrounding areas) to the ozone air concentrations during 1999 ozone episode in Madrid (Spain). In: Midgley PM, Reuther M, Williams M (eds) Proc. EUROTRAC-2 Symposium 2000, Springer Verlag, Berlin

San José R, Stohl A, Karatzas K, Bøhler T, James P, Salas I (2001) An extraordinary ozone episode in Madrid (April, 2000) during night time: a modelling study.In: 7th Int. Conf. Harmonization within Atmospheric Dispersion Modelling for Regularoty Purposes. 28-31 May, 2001,Belgirate, Italy, pp 390–394

Silibello C, Catenacci G, Pirovano G (2000) The use of indicator concept to evaluate Ozone-NOx-VOC sensitivity during an ozone episode over Northern Italy. In: Proc 29th International Conference of Automation and Decision Making, pp 509–518

Silibello C, Calori G, Pirovano G, Carmichael GR (2001) Development of STEM-FCM (Flexible Chemical mechanism) modelling system – Chemical mechanisms sensitivity evaluated on a photochemical episode. In: APMS'01 Conference, April 9-13 2001, Paris

Sillman S (1995) The use of $NO_y$, $H_2O_2$ and $HNO_3$ as indicators for ozone-$NO_x$-hydrocarbons sensitivity in urban locations. J Geophys Res 100: 14175–14188

Sillman S (1999) The relation between ozone, NOx and hydrocarbons in urban and polluted rural environments. Atmos Environ 33: 1821–1845

Simpson D (1995) Biogenic emissions in Europe 2. Implications for ozone control strategies. J Geophys Res 100: 22891–22906

Toll I, Baldasano JM (1999) Photochemical Modelling of the Barcelona Area under Weak Pressure Synoptic Conditions. Atmospheric Environment In: Air Pollution VI. Computational Mechanics Publications

Toll I, Baldasano JM (2000) Modelling of photochemical air pollution in the Barcelona area with highly disaggregated anthropogenic and biogenic emissions. Atmos Environ 34: 3069–3084

Veraldi S, Veronesi V, Volta M (2000) Traffic emission abatement strategies: Brescia metropolitan area case study. In: Proc. 29th International Conference of Automation and Decision Making, Milano

Vogel B, Riemer N, Vogel H, Fiedler F (1999) Findings on NOy as an indicator for ozone sensitivity based on numerical simulations. J Geophys Res 104: 3605–3620

Volta M, Finzi G (1999) Evaluation of EU road traffic emission abatement strategies in Northern Italy. EUROTRAC Newsletter 21, 29–34

WHO (1996a) Update and Revision of the WHO Air Quality Guidelines for Europe. Classical Air Pollutants: Ozone and other Photochemical Oxidants. European Centre for Environment and Health, Bilthoven, The Netherlands

WHO (1996b) Update and Revision of the WHO Air Quality Guidelines for Europe. Ecotoxic Effects: Ozone Effects on Vegetation. European Centre for Environment and Health, Bilthoven, The Netherlands

Winner DA, Cass GR, Harley RA (1995) Effect of alternative boundary conditions on predicted ozone control strategy performance: a case study in Los Angeles area. Atmos Environ 29: 3451–3464

URL 8.1: http://aix.meng.auth.gr/saturn/forum/index.htm
URL 8.2: http://www.meto.govt.uk/
URL 8.3: http://loop.web.psi.ch/gerzensee/prot_f.html

# Chapter 9

Bøhler T, Riise A (1997) Using the air quality assessment system AirQUIS in modelling the population's exposure to traffic induced air pollution, NILU F 7/97, Norwegian Institute for Air Research, Kjeller

Bøhler T, Sivertsen B (1998) A modern Air Quality Management system used in Norway, NILU F 4/98, Norwegian Institute for Air Research, Kjeller

Bøhler T (2000) APNEE Report on Evaluation of Field Trials (EU 5th FWP project IST-1999-11517 Delivarable 6.3), available via http://www.apnee.org

Bøhler T, Karatzas K, Peinel G, Rose T, San Jose R (2002) Providing multi-access to environmental data – customizable information services for disseminating urban air quality information in APNEE. Comput Environ Urban 26: 39–61

Daly M (ed) (1998) Good Practice in European Urban Air Quality Management, Sheffield City Council on behalf of the European Commission Directorate General XI. May 1998, Sheffield, UK

Decanini E, Pola M, Polla Mattiot F, Volta M (2002) Application of REMSAD and GAMES modelling systems on a particulate matter and ozone episode in Milan metropolitan area. In: Batcharova E and Syrakov D (eds) Proc. 8th International Conference on Harmonisation within Atmospheric Dispersion Modelling for Regulatory Purposes, Demetra Ltd. Sofia, pp. 350-354

Decanini E, Volta M (2002) Flexible modelling system for air pollution simulation and control in Northern Italy. In: Midgley PM and Reuther M (eds) Proc. EUROTRAC-2 Symposium 2002, Margraf Verlag, Weikersheim

Finzi G, Volta M (2001) A Real Time Urban Ozone Alarm System. In: Proc. 3rd International Conference on Urban Air Quality (CD-ROM edition), 19-23 March 2001, Loutraki

Grønskei KE, Walker SE, Gram F (1993) Evaluation of a model for hourly spatial concentrations distributions. Atmos Environ 27B: 105–120

Karatzas K (2000) Preservation of environmental characteristics as witnessed in classic and modern literature: the case of Greece. Sci Total Environ 257(2-3): 213-218

Karatzas K (2002) Theoretical Investigation of Urban Air Quality Management Systems Performance Towards Simplified Strategic Environmental Planning. Water Air Soil Poll, Focus 2, No 5-6: 669-676

Karatzas K, Moussiopoulos N (2000) Development and Use of Integrated Air Quality Management Tools in Urban Areas with the aid of Environmental Telematics. Environ Monit Assess 65: 451–458

Karatzas K, Moussiopoulos N (2000) Urban air quality management and information systems in Europe: legal framework and information access. J Environ Asses Pol Manag 2, No 2: 263–272

Karatzas K, Papadopoulos A, Moussiopoulos N, Kalognomou E, Bassoukos A (2000) Development of a hierarchical system for the teletransmisssion of environmental and energy data. Telematics and Informatics 17: 239–249

Karatzas K, Moussiopoulos N, Fedra K, Lohmeyer A, Kouroumlis Ch (2001) A multimedia application for EIA studies. IEEE Multimedia 8, No 4: 71–75

Karatzas K, Moussiopoulos N, Papadopoulos A (2001) Web-based tools for environmental management, Environmental Management and Health 12(4): 356–363

Karatzas K, Dioudi E, Moussiopoulos N (2002) Identification of major components for integrated urban air quality management and information systems via user requirements prioritisation. Environ Modell Softw, in press

Karppinen A, Kukkonen J, Elolähde T, Konttinen M, Koskentalo T, Rantakrans E (2000a) A modelling system for predicting urban air pollution, Model description and applications in the Helsinki metropolitan area. Atmos Environ 34: 3723–3733

Karppinen A, Kukkonen J, Elolähde T, Konttinen M, Koskentalo T (2000b) A modelling system for predicting urban air pollution, Comparison of model predictions with the data of an urban measurement network. Atmos Environ 34: 3735–3743

Karppinen A, Härkönen J, Kukkonen J, Aarnio P, Koskentalo T (2002) A statistical model for assessment of regionally and long-range transported proportion of urban PM2.5. Scand J Work Environ Health, in print

Kukkonen J, Valkonen E, Walden J, Koskentalo T, Aarnio P, Karppinen A, Berkowicz R, Kartastenpää R (2001a) A measurement campaign in a street canyon in Helsinki and comparison of results with predictions of the OSPM model. Atmos Environ 35: 231–243

Kukkonen J, Härkönen J, Walden J, Karppinen A, Lusa K (2001b) Evaluation of the dispersion model CAR-FMI against data from a measurement campaign near a major road. Atmos Environ 35: 949–960

Kukkonen J, Härkönen J, Karppinen A, Pohjola M, Pietarila H, Koskentalo T (2001c) A semi-empirical model for urban PM10 concentrations, and its evaluation against data from an urban measurement network. Atmos Environ 35: 4433–4442

292

Kousa A, Kukkonen J, Karppinen A, Aarnio P, Koskentalo T (2001) Statistical and diagnostic evaluation of a new-generation urban dispersion modelling system against an extensive dataset in the Helsinki Area. Atmos Environ 35: 4617–4628

Kousa A, Kukkonen J, Karppinen A, Aarnio P, Koskentalo T (2002) A model for evaluating the population exposure to ambient air pollution in an urban area. Atmos Environ 36: 2109–2119

Larssen S, Grønskei KE, Gram F, Hagen LO, Walker SE (1994) Verification of urban scale time-dependent dispersion model with sub-grid elements in Oslo, Norway. Air Poll. Modelling and Its Appl. X, Plenum Press, New York

Lohmeyer A, Flassak Th, Fedra K, Karatzas K, Prattos G, Moussiopoulos N (2001) AIR-EIA - Informationsquelle zur Umweltverträglichkeitsuntersuchung für das Schutzgut Luft in Deutschland, Griechenland und Österreich, NDF Information Wissenschaft und Praxis 52(4): 211-215

Öttl D, Kukkonen J, Almbauer RA, Sturm PJ, Pohjola M, Härkönen J (2001) Evaluation of a Gaussian and a Lagrangian model against a roadside dataset, with focus on low wind speed conditions. Atmos Environ 35: 2123–2132

Sivertsen B, Bøhler T (2000) On-line Air Quality Management System for Urban Areas in Norway, NILU F4/2000, Norwegian Institute for Air Research, Kjeller

Slini Th, Karatzas K, Moussiopoulos N (2002) Statistical analysis of environmental data as the basis of forecasting: and air quality application, Sci Total Environ 288: 227–237

Slørdal LH (2000) Wood burning and suspended particulate matter, NILU Draft Report, Norwegian Institute for Air Research, Kjeller (In Norwegian)

Slørdal LH (2001) Applying model calculations to estimate future urban air quality with respect to the requirements of the EU directives on NO2 and PM10. Presented at the 2nd International Conference on Air Pollution Modelling and Simulation, Paris 9-13 April, 2001. (To appear in a Springer publication in their Environmental Science and Geoscience Program)

Tiitta P, Raunemaa T, Tissari J, Yli-Tuomi T, Leskinen A, Kukkonen J, Härkönen J, Karppinen A (2002) Measurements and Modelling of PM2.5 Concentrations Near a Major Road in Kuopio, Finland. Atmos Environ 36: 4057–4068

URL 9.1: http://www.apnee.com

# Subject Index